LEAN PROJECT MANAGEMENT

Many organisations face the challenge of making their project management more agile. However, the circumstances are often not suitable for this: The desired agility either does not fit the existing projects, or there is a lack of sufficient systematics. *Lean Project Management* shows how the advantages of different Lean Project Management methods – adaptive, targeted and flexible – can be combined. In addition to the established methods of classic and agile project management, proven methods and tools from Lean Management are used and further developed with a view to the requirements of project management (such as Gemba, 5S and more). The book shows how an organisation can systematically professionalise its project management, and make it more flexible in a targeted manner, in order to achieve more value with less effort. Aimed at students on postgraduate courses in business and project management as well as professional project managers working in organisations both large and small, *Lean Project Management* is a clear and comprehensive guide to combining the best methods to achieve optimal results.

Claus Hüsselmann is head of the Laboratory for Process and Project Management in the Department of Industrial Engineering at the TH Mittelhessen University of Applied Sciences, Germany.

LEAN PROJECT MANAGEMENT

Claus Hüsselmann

Routledge
Taylor & Francis Group

LONDON AND NEW YORK

Designed cover image: Getty © Olivier Le Moal

First published 2024
by Routledge
4 Park Square, Milton Park, Abingdon, Oxon OX14 4RN

and by Routledge
605 Third Avenue, New York, NY 10158

Routledge is an imprint of the Taylor & Francis Group, an informa business

© 2024 Claus Hüsselmann

British Library Cataloguing-in-Publication Data
A catalogue record for this book is available from the British Library

Library of Congress Cataloging-in-Publication Data
Names: Hüsselmann, Claus, author.
Title: Lean project management / Claus Hüsselmann.
Description: Abingdon, Oxon ; New York, NY : Routledge, 2024. |
Includes bibliographical references and index.
Identifiers: LCCN 2023024773 | ISBN 9781032556543 (hardback) |
ISBN 9781032556468 (paperback) | ISBN 9781003435402 (ebook)
Subjects: LCSH: Project management.
Classification: LCC HD69.P75 H87 2024 | DDC 658.4/04–dc23/eng/20230525
LC record available at https://lccn.loc.gov/2023024773

ISBN: 9781032556543 (hbk)
ISBN: 9781032556468 (pbk)
ISBN: 9781003435402 (ebk)

DOI: 10.4324/9781003435402

Typeset in Times New Roman
by Newgen Publishing UK

CONTENTS

FOREWORD

In the middle of the last century, Toyota developed management principles that were instrumental in helping the Japanese car industry to catch up with Western car manufacturers, and in many cases to outperform them.

Western car manufacturers had relied for too long on their supposed economies of scale and a supposedly secure market situation with stable demand. Processes and planning were unable to respond to short-term fluctuations. Long-term detailed plans that left little room for change and were often unrealistic led to management methods and products that lagged behind customer requirements, resulting in high costs, errors and waste.

At the end of the 1980s, the opportunities offered by the so-called "Toyota Production System" or lean production were recognised and have since found their way into management systems and literature all over the world.

If we look at the development of project management, parallels become apparent. Project management also had to find its way to systematic and methodical management. Comparable to the transition of the little-structured production not oriented towards economies of scale in the early days of the automobile. As project management became more accepted and professionalised, the planning, management and control tools became more professional, but also more complex and less flexible. As a result, many project management systems lost the necessary pragmatism, but above all flexibility, customer orientation and economic efficiency. It was these shortcomings of "classic" plan-driven project management approaches that paved the way for the great success of "agile methods", which differed significantly from many of the principles of plan-driven project management.

At the same time, our studies on the status quo of agile management in projects since 2012 have repeatedly shown that practitioners rarely act in a purely agile manner, but rather use hybrid or selective project management in the clear majority of cases. In this case, the approaches of classic project management are combined with the approaches of agile management. When analysing the benefits that the agile impulses are supposed to bring to project management, there is a significant congruence with the goals of lean management, for example in the emphasis on customer orientation. It is more than a coincidence that almost all concepts and also the masterminds of agile methods refer very strongly to Lean principles – regardless of whether this is at team level or at a higher level.

This book by Claus Hüsselmann takes this into account and takes the necessary next step for project management. Lean PM creates a regulatory framework that aims to apply and combine methods to add value, regardless of whether they are "classic" or "agile". The basic principles of Lean Thinking are applied to projects, but also concrete practices for their implementation are presented. Valuable guidance for choosing the appropriate project approach is provided in particular by the systematic criteria catalogues for characterising the project at hand and its environmental conditions. This provides users with a variety of well-founded tools for the targeted design of hybrid methods – and thus supports readers in taking the next step in the further development of project management.

Prof. Dr Ayelt Komus, Koblenz University of Applied Sciences

PREFACE

My first project in a professional environment was the development of an environmental information system in an integrated SAP platform in the mid-1990s. Shortly after my studies, I was allowed to lead this project as primus inter pares of a four-member team. The result was the first of a series of SAP-based specialist applications, which was then also put into production at the environmental authority that commissioned it. During this time, the idea of process-oriented design as well as the standardisation of project management took off significantly. My passion for these domains was ignited with the successful project.

This was followed by a large number of other small and large organisational and IT projects that I was allowed to carry out and also lead. I report on the two largest ones in this book. They provide an essential practical background for the ideas of Lean Project Management. As early as the end of the 2000s, my team of experienced project managers developed the concept of a hybrid approach for SAP implementation projects to utilise the respective advantages of the so-called waterfall and agile approaches. During this time, I was responsible for the *Project Operations & Risk Control* unit of a globally operating TecDAX company. Here, as a standardiser, coach, trainer and supervisor of the company's A and B projects, I gained further valuable insights and suggestions. The idea of Lean Project Management as a framework for the promising contextual design of project procedures began to take shape.

With my move into the world of research & teaching in 2015, I was able begin to systematise the ideas. This book is the preliminary result of that work. It is intended to help readers and their organisations to give a methodological framework to ways of thinking and practices that can certainly also be implemented intuitively with the much-invoked common sense, and to establish them in the user organisations and their projects.

I hope you enjoy reading it and have a few "aha" moments.

Claus Hüsselmann

ACKNOWLEDGEMENT

Many of the Lean PM practices presented are the result of my own reflections and experiences, and some have been adopted from other authors and, where appropriate, slightly modified. No idea, however, has been made possible or developed without a professional discourse with companions. My thanks go to them!

I would especially like to thank Professors Bert Leyendecker, Ayelt Komus, Matthias Vieth, Dorothee Feldmüller, colleagues Markus Götz, Sonja Schmidt, Lisa Rost, Claus-Peter Koch and Rüdiger Kloss for their cooperation and valuable input in developing the topics of Lean PM.

Furthermore, thanks also go to the students of the department of Industrial Engineering who contributed to the development. The work of Maximilian Heymann and Paul Golfels, Niklas Haemer, Henrik Wilhelm, Phillip Baumann, Arber Haliti and Selim Kan deserves special mention.

Last but not least, I would like to thank Frank Baumgärtner, Traudl Kupfer, Andrew Harrison and Alison Craig for their professional and constructive critical support of the book project in the German origin and this English-language edition of the book.

"Lean is a journey, not a destination!" ... following these words of the Lean Construction pioneer Glenn Ballard, the development of Lean PM, Lean-Agile PPM and UPMF continues. Interested readers are welcome to participate and contact me (www.lean-projectmanagement.de).

SUMMARIES

Introduction

In this chapter we discuss the critical importance of projects and thus of project management for economic value creation. Megatrends such as digitalisation and globalisation have a particular influence here because they reinforce the necessity of projects as a form of work, but they also place specific new demands on them.

Although project management methods have been further developed over many decades, success rates still leave much to be desired. This applies not least to the world of IT projects with their fast pace, and has led to a paradigm shift in this domain – from a plan-driven to an agile approach.

In order to meet the requirements of modern project management, which can be characterised by flexibility, lightness, practicability but also universality (not all projects are IT projects), it is necessary, according to the author, to "stretch an umbrella over the different approaches", so to speak, which makes it possible to select the optimal approach specifically depending on the context of a project. This has motivated the development of Lean Project Management.

Chapter 1

This chapter serves to introduce the reader to the two core topics of the book – project management and Lean Management. For this purpose, the key terms of both domains are described in a definitional way.

In summary, a project is an activity that has a unique character in its entirety, pursues a clear goal, is limited in time and has a certain organisational and technical complexity. The management of projects aims to achieve the defined objectives within the given constraints. The aim is to achieve a balanced intensity of project management that promotes the performance of the project team – as neither too much nor too little is conducive to achieving the goal.

In terms of methodology, project management must be distinguished from the technical and progressive development of project products on the one hand and the higher-level project portfolio management on the other. For the former, the chapter briefly touches on some typical process models, such as the waterfall or spiral model. For project management, on the other hand, a general reference model is presented, the so-called Unified Project Management Framework, which has been developed at the University of Applied Sciences Centralhesse (GER). This is universal because it generally postulates a procedure according to the Plan-Do-Check-Act cycle and defines ten central sub-disciplines of PM.

With regard to Lean Management, this chapter describes its roots in the production sector. Following the central paradigm of avoiding waste, the types of waste known from production are briefly characterised. In particular, the core principles of Lean Management according to Womack and Jones are discussed: Customer and Value Stream orientation, Flow, Pull and Perfection, as these are to be adapted for Lean Project Management in the following chapters.

Chapter 2

Agile approaches can be characterised as manifestations of Lean Management. This chapter therefore defines the term and examines the underlying principles and objectives. Two sources in particular are examined – the Agile Manifesto and the Scrum process model, both of which originate from the field of software development. As with Lean Management, a central paradigm and core principles can be identified for Agility: Flexibility and adaptability (paradigm), and collaboration, delegation, learning, iteration and simplicity (core principles). These also have to be operationalised through appropriate practices.

So-called hybrid approaches have become established in project practice, and try to derive their own approach from different approaches (especially plan-driven and agile). Different modus operandi can be observed here, which will be dealt with methodically, but also with practical examples. This also motivates the following detailed analysis of Scrum with regard to its completeness as a general PM approach. As a result, some gaps become apparent, due to Scrum's origin in the organisation of software development.

A further focus of the chapter is the analysis and systematisation of the concept of complexity, since – similarly to agility – an almost inflationary use of this concept can be observed. Here, too, the first step is to define the term, focusing in particular on the difference between complex and complicated systems. This is followed by a description of selected standard strategies for dealing with complexity in order to resolve the identified dilemma in the application of agile versus plan-driven approaches.

Chapter 3

After the introductory remarks of the first chapters, this chapter finally derives and describes the basic concept of Lean Project Management. In addition to a definition, the core principles and terms of Lean Management are interpreted in the context of project management (PM). This includes, in particular, the question of what waste actually is in the context of projects or project management, since in many projects, in contrast to production, it relates to intangible value streams, such as the flow of information. PM-specific types of waste are identified as Waiting, Over-processing, Defects, Misallocation, Misdirection, unnecessary Movement and Under-processing.

All the other core principles of Lean Management – customer focus, the flow and pull principles, but also the pursuit of perfection – require a specific interpretation. The pull principle proves to be more in need of adaptation, but nevertheless promises innovative practices, such as work-in-progress limitation, applied through a project Kanban board. Last but not least, the idea of Lean Project Management leads to a reinterpretation of the well-known Magic Triangle of PM, which induces in particular the customer- and thus application- and benefit-orientation of project results.

In the form of three simple guidelines (the "3Gs") – "Participation", "Application of the Pareto Principle" and finally "Fit" – guidance is finally formulated on how to design projects or project management according to Lean Project Thinking. These are underpinned by a number of application elements, such as the involvement of "downstream" stakeholders in "upstream" activities.

Chapter 4

This chapter is a key chapter in the systematic operationalisation of Lean Project Management because it describes a variety of practices for implementing the core principles. Practices in this context are action principles such as the Gemba ("go local") or the demand for standardisation, but also concrete methods and tools. While the implementation of the core values of Lean (Project) Management is essential (focus on value creation for the customer, etc.), the use of practices is to be understood as a possible aid to achieve this implementation. The selection in this chapter is certainly comprehensive but cannot be described as complete, since the latter seems hardly possible – as anything that helps is allowed.

The practices presented are classified according to their areas of application: Contract Design, Scope Management, Project Planning, Process-Oriented Control, Continuous Improvement ... just to name the umbrella terms. Many of these practices have their roots in Lean Software Development and Lean Construction, but they are all described independently so that they can be used in any type of project. Some of the practices have also been newly developed in the course of shaping the concept of Lean Project Management. One of these is the so-called Agilometer, which makes it possible to systematically record a project's boundary conditions and deduce whether a plan-driven or an agile approach is more appropriate. This is a question faced by many users in the PM domains of companies.

This central chapter concludes with the application of value stream mapping to project management. This is done by first identifying the value streams and then orchestrating them end-to-end, using the processes of the Unified Project Management Framework as examples for project risk management and project knowledge management. For each of these value streams, a detailed description is given of who the process customers are, what waste typically occurs, and what core principles and practices can be used to reduce it. The direction is given by the formulation of "Benefits Expectation Stories" which are developed within the framework of the concept – a sentence template that can be used to formulate the expectations of process customers in a lean and consistent manner.

Chapter 5

This chapter describes two "flagship" projects from my own experience as a project manager, which have contributed significantly to the development of the Lean Project Management concept from the bottom up, so to speak. The first was an SAP implementation in a corporate structure with about 50 legal entities. This project was significantly characterised by the requirement for cross-organisational harmonisation of the HR processes of independent authorities in many different locations, the replacement of a legacy system, the corporate cultural challenges faced by those involved and, last but not least, technical complexity.

The second project involved the first-time establishment of a central control centre for emergency call processing and dispatching of emergency forces in a federal state that could be used throughout the state. The challenges here were not least of a technical nature, but the existing state political and media attention should also be mentioned. Risk management required particular attention, as there were a number of important influencing factors, many of which were outside of the project's own sphere of influence, e.g. the provision of a country-wide Digital Mobile Radio network for Emergency Services. In this project, a "third party approach" came into play, in which (as is usually the case only in construction projects) a project "operator" (aka controller) acted as project manager for the PM processes and the PM system – alongside the client's project manager, comparable to a product owner.

In both projects, methods from agile approaches were successfully applied according to the situation, with the basic orientation in each case being plan-driven. In the end, hybrid approaches emerged that followed the strict benefits orientation familiar from lean thinking. The chapter presents a variety of these methods.

Chapter 6

This chapter presents a canon of criteria that makes it possible to derive the requirements for the PM system of a project *a priori* on the basis of its characteristics. This implements the premise that "approach follows context", which expresses that each project has management requirements according to its individual context. For example, the construction of a production hall is unlikely to benefit from fully developed organisational change management, whereas this is essential for a reorganisation project.

In the style of a meta-analysis, a morphological scheme has been developed from various sources. With this scheme, 18 criteria and their minimum to maximum values can be used to determine the characteristic profile of a given project. Typical criteria are complexity, urgency and degree of innovation.

If this profile shows specific focal points, then these in turn can be used to derive the necessary design focal points for project management, which are also described. The project example of the creation of an audio tour for a museum rounds off the profiling for illustrative purposes.

In summary, it can be said that the requirements for the alignment of the management of a project are determined 1st by the type of project (e.g. construction project versus IT project), 2nd by the internal project category (e.g. A, B or C projects) and finally 3rd by the individual project profile as described.

Chapter 7

Lean Project Management is a concept that is both at the level of mindset (Lean Thinking) and at the level of concrete practices (methods, tools … "recipes"). The implementation or application cannot therefore generally be done "at the push of a button" but requires the planned action of a change project. As we know from change management, people need to be brought along on the journey to the new if the journey is to be successful and bring benefits.

However, in the context of project management, it should be noted that not all people or corporate cultures are suitable for every management approach. For example, some people only develop their potential with complete freedom of design, while others much prefer to follow clear instructions as implementers. Consequently, these almost unchangeable (because human, character-related) boundary conditions must be taken into account when designing an organisation's PM system. Based on some well-known sociological concepts (by Hofstede, Hall, Lewis and others), a connection with project business is derived in this chapter. Finally, the St. Gallen Management Model provides a framework for the question of which management functions should be sensibly fulfilled in the team or through the leadership functions. Independent of all the sociological classification models described, the insight is justified that it is not the character of the manager but the circumstances in the team that should induce the required management style on a case-by-case basis.

The second part of the chapter deals more with the "hard facts" of how to implement Lean Project Management at the three levels of designing the corporate PM system, proactively setting up the PM system of an individual project, and situational, more reactive application during the course of a project. Various approaches, guiding questions, practices and key figures are presented. Finally, success factors are listed which, if taken into account, will improve the chances of success.

Chapter 8

Lean Project Management, as presented in this book, focuses primarily on the level of individual projects. In the reality of companies, however, it is not just individual projects that are carried out, but often a multitude of projects at the same time. From the interaction of these projects – e.g. in the "battle" for scarce resources – further aspects of (higher-level) control arise, which project portfolio management (PPM) deals with.

With the observation that the process and control mechanisms in agile project approaches differ significantly from the plan-driven, often hierarchically structured ones, there are also implications at the multi-project level. The increasing dynamism of the corporate environment also leads to changed requirements, e.g. for the duration of projects and thus the time horizon of the composition of the project portfolio.

This motivates the outlook on the idea of Lean-Agile Project Portfolio Management, in which the principles and practices of agility and lean thinking are "lifted" to the multi-project level in a comparable way. This chapter outlines approaches to this, the details of which, however, go beyond the scope of this book.

Nevertheless, the central elements of these approaches, such as flexibility, customer orientation, value stream identification, flow and pull principles, etc., are initially transferred to the project portfolio level and applied – comparable to the procedure at the individual project level. Based on criticisms of the classic, stability-oriented PPM and empirical success factors for PPM, suggestions are given for operationalising lean-agile design principles for PPM. In addition, a methodical extension of the well-known RACI matrix is applied as an example for PPM, in which customer benefit orientation is recorded.

INTRODUCTION

After reading this chapter, you will know ...

- what the challenges of project management are in today's world, which is characterised as a "VUCA" world, and
- what the motivation is for Lean Project Management as an overarching concept, covering plan-driven and agile approaches.

Challenges in modern project management

The leading study in Germany (by the GPM) on the macroeconomic measurement of project activity shows that the share of project activity in total working time was about 35% across Germany in 2013/2014. And the "projectification" of the economy, i.e. the increase in project work in companies, will continue.[1] Increasingly more value creation is taking place in the form of projects. Let us take a closer look at two general trends in economic life that have a direct influence on the subject under consideration – the project economy – globalisation and digitalisation.

Globalisation in the project economy means increased international cooperation, and thus collaboration, under the framework conditions of distributed locations and different (work) cultures. Additionally, project teams are becoming more unstable, or in other words: More employees are involved, may be replaced in the course of a project, or may be added to a (major) project later, etc. For example, in a multi-year global SAP consolidation and roll-out project in a pharmaceutical company, the project contractors were replaced several times due to procurement regulations and likewise, own employees took on different functions in the project and in the company over time. This is all in the nature of large-scale projects. However, it is obvious that this results in waste due to increased familiarisation time and know-how transfer losses.

---

1 cf. GPM 2015, p. 19 ff.

DOI: 10.4324/9781003435402-1

The second megatrend, digitalisation, currently seems to dominate attention in the economy. Gartner (2018) found that artificial intelligence, the digitalisation of business processes and models, and networking (and the individual developments that can be subsumed under it) currently determine the technological trends.[2] Digitalisation means IT-driven innovation! And the development of IT is not linear and is proceeding at an unprecedented speed. According to *Moore's Law*, the capacity of digital integrated circuits doubles in a period of about 18 months. No matter which interpretation of this law one chooses, the realisation of a non-linear increase in the performance of IT remains, and has been confirmed in the past decades, which everyone can see in everyday life. Just think of the performance (and importance) of smart phones, which first appeared in 2007 with the iPhone 1. In addition, the networking of people and things through internet technology is also developing rapidly. In 2015, for example, about 25 billion things were connected to the internet – this number had been doubled to over 50 billion by 2020.[3] The increasing possibilities of IT are massively driving technical and business developments. In many cases, they offer relatively low barriers to market entry (an app is quickly programmed, see Uber and others) and enable continuous development of products even after market launch (see, for example, updates at Amazon every second),[4] and thus lay the foundation for an ever greater speed at which products and services are developed, provided and further developed.

The digital transformation of companies also means, however, that companies whose core competence to date has been, for example, the (conventional) production of mechanical engineering parts, suddenly find themselves confronted with an extensive and subsequently business-critical need for IT, the development of which also exhibits the aforementioned dynamics. If the development of mechatronics has already led to a significant increase in product and development complexity, digitalisation adds another dimension to this. The result: The complexity of processes and projects in the company increases significantly; and dealing with uncertainty becomes a central management issue.

It is also noted that the corporate project landscape is characterised by an extraordinary diversity of projects. In a leading study of the Technical University of Berlin, for example, an average of about 120 projects were managed in the project portfolios investigated (n=200).[5] Another study by GPM also shows that about two thirds of the Project Management Offices (PMO) manage portfolios with a wide variety of project types (IT, R&D, organisation, investment).[6]

In the course of these developments, and from the bitter realisation of the still poor general success rates of projects,[7] agile and classic approaches have formed poles of project management, so to speak. Classical means the established international best practices and standards (from PMI, IPMA, PRINCE2, ISO/DIN etc.).[8] These can be characterised as plan-driven, i.e. they postulate a systematic planning of projects and project phases in many respects (structure, process, quality assurance, risk management, etc.). Contrary to what is often said,

---

2 cf. Schmitz 2018.
3 cf. Robinson 2015.
4 see McKendrick 2015.
5 see Gemünden et al. 2011.
6 see GPM 2014.
7 cf. e.g. PMI 2018, p. 14.
8 PMI: Project Management Institute; IPMA: International Project Management Association; PRINCE2: Projects in Controlled Environments, Vers. 2; ISO: International Organisation for Standardisation; DIN: German Institute for Standardisation.

this does not automatically imply the so-called *waterfall approach.* Agile approaches generally operationalise the requirements of the Agile Manifesto from 2001.[9] The approach is always iterative and incremental and can be characterised as fundamental to, and largely adaptable to, changes during the project. In many cases, upstream planning is subordinate or even rejected and the self-organisation of the team is emphasised (cf. Scrum).

This polarisation creates confusion and irritation, especially for organisations that do not practice project management (PM) in their core business, for example small and medium-sized industrial companies. Here, the experience of many supervised student projects in companies shows that small and medium-sized enterprises (SMEs) have not yet reached an adequate level of maturity in PM but are already caught up in the wave of agility. In general, difficulty also arises in handling the poles of PM, both of which have their contextual justification, in one management system. The management of multimodal project landscapes also poses challenges for established portfolio managers (methodologically, organisationally and procedurally).

So, we live in a *VUCA-world: Volatility, uncertainty, complexity and ambiguity* which creates significant demands on the management in control of companies. A number of national and global trends and megatrends have emerged[10] that characterise and cause VUCA – see the examples of digitalisation and globalisation.

The following requirements, which are basically not new, can be derived for modern PM in the VUCA-world:

- **Flexibility**
  A modern PM system must be flexibly adaptable to the respective requirements of the projects and the organisation. Depending on the project context, for example, a plan-driven approach (e.g. expansion of a production hall) or an agile approach (e.g. development of a new, internet-based service) may be appropriate.
- **Lightweight**
  Small projects do not want to shoot cannons at sparrows and must avoid administrative overhead. Employees must quickly find their way around the PM system.
- **Practicability/practical orientation**
  Although a comprehensive theoretical foundation is helpful for the project management of an organisation as a whole, the practitioners, i.e. the project managers in the field, demand simple, clear and goal-oriented recipes for managing projects.
- **Universality**
  The PM system of an organisation must be applicable to all existing project types. A portfolio management system in which Scrum-projects report according to method X and infrastructure projects according to method Y is not expedient.

The application of Lean Thinking, adapting the various forms of Lean Management for production, administration, product and, last but not least, IT development, to project management promises

---

9 see Beck et al. 2001.
10 see Scheller 2017, p. 16 ff.

to fulfil the requirements of modern PM. Lean PM is characterised by a high degree of customer and value-added orientation while avoiding waste (especially administrative overhead) as far as possible. Methods and tools known from Lean Management are used and further developed with a view to the requirements of project management, for example Gemba, 5S and others (see Chapter 5).

Motivation for Lean Project Management

PM is currently undergoing a (supposed) paradigm shift: From a linear, sequential approach (waterfall) to a cyclical, incremental one (agility). In the advancement of the science and above all practice, hybrid approaches are increasingly emerging that combine the best of both worlds.[11] It can also be seen that the apparently opposing approaches (may) have many things in common when inspected closely – even if they undoubtedly emphasise and highlight different planning and control philosophies and offer corresponding methods (e.g. Scrum).

Was everything bad in the past? This question must inevitably be asked if one follows the current interest in the topic of agile projects. Many empirical studies – above all the widely cited CHAOS report by the Standish Group – allow this conclusion. After all, according to their own information, 40,000 to 50,000 (IT) projects have been examined since 1994.[12] The 2015 report states that only 11% of the more than 10,000 projects examined were successful, the rest were not within the target corridor in terms of deadlines, service delivery and/or costs, or even failed completely.[13] The situation is different with agile projects: There, the success rate is 39%. Furthermore, a correlation with the size of the projects is shown: Regardless of the process model, larger projects fail more often than smaller ones. Lighthouse projects, such as the much-cited Berlin Brandenburg Airport, are not really evidence of the successful maturity of (large) projects.

The simple conclusion to be drawn from the above would be that projects that are implemented according to the classic models such as the PMI, the IPMA or PRINCE2 would have structurally worse chances of success. But what if the application of the "old" PM methods was simply not carried out in a target-oriented manner in many cases? After all, excellent projects in Germany (GPM) and worldwide (IPMA, PMI) show that projects can be carried out successfully and "with a plan".[14]

And can agility be transferred so easily to construction projects, for example (see Figure I.1)?

The most commonly used method in agile is Scrum.[15] Scrum has its roots in IT development projects and in its original form assumes a team size of about seven employees. Such a project therefore generally belongs to the small ones.[16] In this respect, attempts have emerged in recent years to scale the Scrum approach to larger project contexts, which are common in the practice of large organisations. These include the *Scaled Agile Framework* (SAFe), *Large-Scale Scrum*

---

11 cf. Komus/Kuberg 2017, p. 26.
12 cf. The Standish Group 2018.
13 cf. Wojewoda/Hastle 2015.
14 see GPM n.d.; IPMA n.d.; PMI n.d.
15 cf. Komus/Kuberg 2017, p. 13.
16 cf. Schwaber/Sutherland 2017, p. 6.

FIGURE I.1 Agile house building?

(LeSS) or *Nexus*.[17] The transferability to other types of projects, such as hardware product development, is also increasingly being investigated.[18]

Certainly, the success story of Scrum is in part due to its conceptual lightness. After all, the original script by Schwaber and Sutherland fits on 17 pages (!) in the current version.[19] Other works, which do not so much describe a concrete method, but rather a *body of knowledge,* come to many hundreds to thousands of pages. These include, of course, the PMI's *Project Management Body of Knowledge*, but also, for example, the *Handbook for Practice and Continuing Education in Project Management (PM4)* of the GPM.[20] The current figures for the low-threshold *PRINCE2*-certifications, amount to approx. 1.2 million certified worldwide by 2018: Practice demands lighter-footed PM systems.[21]

Especially for SMEs the flagships of PM often seem too heavyweight and therefore do not find widespread distribution for a long time.[22] This is when the lean approaches of agile come in just at the right time, and can be recognised in many areas as a concrete application of lean thinking (even if the term does not appear in Schwaber/Sutherland). However, it is easy to see that the well-known agile methods, which in addition to Scrum often include Xtreme Programming and the use of Kanban,[23] do not represent a fully-fledged PM system, as these concepts largely lack essential disciplines such as risk, stakeholder or contract management. Last but not least, Schwaber and Sutherland describe Scrum as a "process framework for managing work on complex products" – and not as a PM method.[24]

17 see SAFe 2020; LeSS n.d.; Schwaber/Sutherland 2018.
18 see Komus 2018, p. 11.
19 see Schwaber/Sutherland 2017.
20 see PMI 2017; GPM 2019.
21 see Brecht-Hadraschek 2014, p. 9; Note: PRINCE2 is not to be classified as a body of knowledge, but rather as a concretely designed PM framework.
22 cf. Vogelsang/Olberding 2007, p. 1.
23 see Bowes 2015.
24 Schwaber/Sutherland 2017, p. 4.

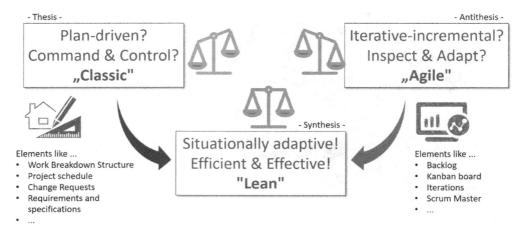

FIGURE I.2 The dialectic of PMs.

A contemporary PM approach should generalise the best of these worlds, and unite and operationalise them in a more universal approach – a *Lean Project Management approach*. Lean PM creates the synthesis in the sense of a modern, goal-oriented and flexible PM approach (see Figure I.2).

The claim of the Lean PM approach – the term was first mentioned as early as 2003 in a paper by Ballard[25] – is to deliver a fully-fledged PM system that enables projects to deliver more value, in terms of project benefits, with less effort. For example, studies of aerospace projects have shown that in the projects studied, the purely value-adding activities accounted for only about 12–13% of the activities![26]

Lean PM ...

- helps to bring projects quickly to the finish line and to stay within project budgets by focusing on the essentials and avoiding the unnecessary (*waste*).
- focuses on the customer's wishes. The customer defines value and value creation. Customer wishes are translated into concrete, measurable *critical-to-quality requirements* for everyday project work.
- is resource-saving because waste and the unnecessary are avoided.
- puts all requirements for documentation, formalities, quality gates, etc. to the test and measures them individually according to what added value they deliver. Anything that does not add value will not be done.

---

25 see Ballard 2002.
26 see Belvedere 2019, pp. 411–413.

Lean PM aims to further develop existing approaches to modern PM and to formulate a generally applicable, lightweight approach that is easily adaptable and open to flexible application between *planning certainty* on the one hand and *trial & error* (in a positive sense, *inspect & adapt*) on the other. Having a (sensible) plan is not fundamentally seen as a bad thing (see also Figure I.1). Lean PM provides an independent, superordinate model that is suitable for current applications, but also for future ones.[27]

27 see also Erne 2019.

1

BASICS FOR LEAN PROJECT MANAGEMENT

After reading this chapter, you will know …

- the relevant basics of projects and project management by definition,
- the big picture of the delineation of the technical project approach, project management and project portfolio management,
- an overview of the so-called Unified Project Management Framework as a PM reference model and regulatory framework for the design of Lean Project Management, and
- relevant basics of Lean Management including its core principles.

1.1 Project management

1.1.1 Value creation through projects

In fact, projects have always existed. The Old Testament already tells of the construction of the Tower of Babel, which came to a standstill due to insurmountable communication difficulties (triggered by a higher authority). The building of the pyramids is also often cited as an example of original projects from the third millennium BC. Even then there were successful and less successful projects. For example, the famous crooked pyramid of the Pharaoh Sneferu was obviously completed in a way that deviated from the original design – apparently due to construction problems – but was nevertheless used as a place of worship. Parallels to today's building projects can be quickly drawn here. Of course, we can assume that the methods of planning and controlling ancient projects were not comparable to those of today's projects. Especially the resource of labour may have been handled differently in the ancient projects – even if today some may still speak of "project slaves". In the further course of history, there have always been large, complex projects of a unique kind, such as the construction of the Great

DOI: 10.4324/9781003435402-2

Wall of China, the Panama and Suez Canals, the Eiffel Tower and the establishment of war fleets, to name but a few. All these projects undoubtedly required extensive planning as well as logistical and control measures – in short, an effective management.[1]

Although Gantt developed his famous bar chart in the early 20th century, marking the first documented PM method, the systematic development of PM is generally associated primarily with the military and civil aerospace programmes after the Second World War. These include, above all, the realisation of the Polaris programme of the US NAVY, the major programmes of the US Air Force and the Apollo programme of NASA, i.e. all US-based projects that gave rise to more comprehensive (project) management systems such as *Phased Project Planning* (NASA) or the *System Program Management* (USAF). With the *Program Appraisal & Review Technique* (PERT) and the *Critical Path Method* (CPM), which were also developed in the USA, the network planning technique then became virtually synonymous with PM in Germany for many years.[2]

Not least in the course of the spread of these approaches to other branches of industry, in particular the automotive industry, the still leading organisations for PM worldwide were founded from the late 1960s onwards: The Project Management Institute (PMI) in the USA in 1969, the International Project Management Association (IPMA) in 1965, initially headquartered in Zurich, or for Germany the German Association for Project Management (GPM) in 1979 as the IPMA national association. The spread of these approaches and formation of these organisations was also associated with the further development of PM as a holistic management approach – complementing the rather technocratic views towards project management with a more appropriate emphasis on social competences, as expressed, for example, in the IPMA Competence Baseline (ICB).

The PM organisations were also the ones who promoted the standardisation efforts in PM by publishing their own best practices. The US American ANSI standard 99-001 was created from the PM Body of Knowledge Guide (PMBoK Guide) of the PMI,[3] the DIN 69 900 ff. as the German standard for PM and ISO 21 500 as an international standard – unfortunately, none of them are consistent with each other.

All the institutions and works described above (and others, such as the British Office of Government Commerce, which published PRINCE2) have each developed their own definitions of projects and PM, which will not be listed here. Rather, as a synthesis of these descriptions, this book is based on the following understanding of the terms.

DEFINITION PROJECT

A project is an undertaking with a limited time and cost frame to deliver a set of desired results that – while meeting certain quality requirements – serve to achieve the defined project objectives. It is thus an organisation created for a limited period of time with the purpose of delivering specific results or products in accordance with an overall benefit objective. A project is essentially characterised by the uniqueness of the conditions in their entirety.[4]

---

1 cf. Madauss 2017, pp. 7–9.
2 see Schelle 2013, p. 109 ff.
3 ANSI American National Standards Institute, US American standards organisation.
4 based on PRINCE2 or IPMA ICB and according to DIN 69901.

▶	**Objective -**	Usually a clearly defined end goal
▶	**Complexity -**	Organisational and technical elements from several areas
▶	**Uniqueness -**	No exact repetition of what has gone before
▶	**Uncertainty -**	Characterised by incomplete information and indeterminate developments
▶	**Time limit -**	At the end of the project, the activities are closed
▶	**Life cycle -**	A project has different phases and a defined end
▶	**Planning -**	Economic and technical framework conditions, disposition, design

FIGURE 1.1 Typical characteristics of projects.

The project definition gives rise to the typical project characteristics shown in Figure 1.1.

Each of these characteristics is less or more pronounced – but never equal to zero. Thus, projects are genuinely characterised as (more or less) complex undertakings (see Section 2.3). As a (temporary) operational organisation, the familiar functions of management are thus applied in projects. These include the setting of goals, the development of a strategy for achieving goals, the organisation and coordination of production factors and the management of employees.[5] For projects we can define:

DEFINITION PROJECT MANAGEMENT

Project management is the application of knowledge, skills, tools and methods to project operations to meet project requirements.[6] It is a disciplined process for identifying, coordinating and continuously pooling resources (people, materials, skills, etc.) to meet project/contract objectives within time, cost, resource and quality constraints.

Mintzberg has empirically investigated and described the various roles that a manager generally assumes. These also apply to the management of projects and can also be observed in this context. Accordingly, a (project) manager is a representative, leader, coordinator, information collector, information distributor, informant of external groups, entrepreneur, crisis manager, resource distributor and negotiator.[7] Depending on the interpretation of the role – by the project manager or the basing organisation – different manifestations arise in the exercise of PM. Following the characterisation appropriately formulated by Schäfer, poles can be identified in the application of PM methods towards weak or pronounced manifestations, with the middle

---

5 see Haric n.d.
6 according to PMI.
7 see Mintzberg 1991, pp. 29–35, and Staehle 1991, p. 15.

	Design of the project management		
	too low	*adequate*	*too high*
Role of project manager	Technical driver, technical expert	Methodical driver with professional understanding, entrepreneur, change agent	Executive organ, vicarious agent
Required competence of the project manager	Professional competence is overemphasised, methodological competence neglected	Methodological competence in project and change management, ability to make professional judgements	Methodological competences are overemphasised, strict adherence to regulations and standards of a framework
Positioning of the project in the organisation	Weak, very high dominance of line organisation	Balanced matrix organisation	Strong mixing of project work in line work
Target focus	(Over-)fulfilment of the technical and content-related requirements, time and costs play a subordinate role	Balancing the dimensions (goals) of the Magic Triangle	Time and costs are overemphasised, formal requirements dominate
Degree of formalisation	Project standards are seen as a (necessary) evil	Project standards are necessary, but not an end in themselves, enabling differentiated, beneficial application	Overemphasis on formal processes and standardised results, efficiency decreases
Risks	Intransparency, high (undetected) target achievement risks	Usual project risks are considered a normal part and systematically managed.	Paralysis, demotivation, building of buffers, shifting of risks

FIGURE 1.2   Design of the PM – poles and fitting accuracy

path being the most promising (see Figure 1.2).[8] It is important to find this middle ground when practising effective PM.

The understanding of projects on which this book is based considers them at the levels of PM processes (PM) and technical, progressive project processing (project procedure (PP)) (see Figure 1.3). These are supplemented by a superordinate level of project portfolio management (PPM), which provides the framework but will only be marginally considered in this book (see Chapter 8).

8 see Schäfer 2017, p. 10.

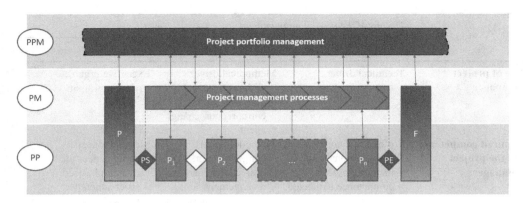

FIGURE 1.3   Project levels (PS: project start, PE: project end).

The professional-progressive approach in a project (PP) depends on the project subject and ultimately on the concrete project context. For example, a construction project naturally has a different underlying procedure model than an IT project, etc. In this respect, the PM level can be classified as more universal than the PP level, which is also expressed in the standards already mentioned. These are generally to be considered independent of the project subject (even if this is often interpreted differently due to the origin of the respective approaches, e.g. PRINCE2 from the IT context).

Nevertheless, different generic models have also emerged for the PM level. The so-called *waterfall model* is the most well-known. In this model, a project is divided into project phases which reflect the fundamental technical sequence of the service provision. It is then assumed that this sequence is also processed sequentially, so that the flow of the creation of the service resembles a waterfall flowing step by step from top to bottom. The model is generally attributed to Royce, who, however, already emphasised that there are iterative feedbacks between the individual phases, which "unfortunately" have to be carried out over several stages in practice.[9] The core idea of Royce and thus of the waterfall model is that there is a logical sequence of activities to develop a product, in which individual technical steps must follow each other according to plan (conception, development, testing, etc.), which must not be skipped if the result is to be usable.

Even in such sequential, plan-driven process models, adaptivity is a prerequisite for stability. This can be illustrated strikingly using the cyberneticist Ashby's example for management: Riding a bicycle in a straight line. If you were to hold on to the handlebars of your bike in this situation, you would inevitably fall over quickly because you would not be able to compensate for the constant small disturbances (fluctuations in balance). A successful PM must therefore – always – be characterised equally by the creation of stability and the preservation of flexibility.[10] Without flexibility, you cannot run the best plan!

Variations of the sequential waterfall model are, for example, the parallel or early processing of technical phases or the so-called *spiral model*. Parallel processing is possible when technical

9 see Royce 1970, p. 330.
10 after Rüegg-Stürm 2005, p. 80.

work can be logically separated from each other or when the availability of partial results already leads to the start of the next logically following work. This has the advantage that work can be started more quickly, but at the price of a higher complexity of the project sequence (incl. risks of rework), which must be managed. The spiral model, on the other hand, basically keeps to the sequential order of the phases, but runs through them several times in the sense of a successive further development of the solution. The number of runs is defined from the outset, which means that the spiral model describes a middle way between purely agile procedures and the waterfall model. In contrast, the purely agile process models do not define *a priori* the number of runs for the creation of a previously clearly defined product but are fundamentally open in this respect – limited only by time and budget conditions.

Which project procedure or PM approach should ultimately be used in a project is the subject of the further explanations of Lean PM in this book, in particular the adaptation of the PM system in Chapter 7.

1.1.2 The Unified Project Management Framework for PM

In the field of (individual) PM, a number of standards and best practices (hereafter simplified as *standards*) have become established. These include in particular the PMBoK Guide (PMI), the ICB (IPMA), Projects in Controlled Environments, PRINCE2 (AXELOS), the Compendium for Competency-based Project Management, PM4 (GPM), DIN 69 900 ff. and ISO 21 500, the process model for IT development projects of the Federal Republic of Germany or the V-Model XT.[11] Due to its importance in the practice of projects, we also include Scrum at this point.

The aforementioned standards all have a common subject of consideration in essential areas – namely the management of projects or product development processes (Scrum) – and differ, among other things, in the technical focus, the wording, the degree of operationalisation or the management philosophy.

Due to the universality of the PM task, one can probably say that it is always the same wine in different bottles. Nevertheless, the standards can basically be divided into so-called *Bodies of Knowledge* (PMBoK Guide, ICB etc.) or operational PM methods/project procedures (PRINCE2, Scrum etc.).[12] Figure 1.4 shows a rough analysis of the different approaches mentioned.[13]

The PM standards are sometimes specifically criticised ("not specific enough", "only applicable to IT projects", "too extensive", etc.) and often divide the PM community into corresponding camps.

Due to the universality of the PM task and the need to be able to develop project-specific methods in the sense of Lean PM (so-called *tailoring*), it makes sense to use or create a universally applicable, scientifically sound and practicable framework. With such a *Unified Project Management Framework* (UPMF, see Figure 1.5)[14] it is then possible to operationalise

---

11 V-Modell XT: Procedure model for IT development projects of the Federal Republic of Germany, "XT" stands for *Extreme Tailoring*.
12 see e.g. Kammerer et al. 2012, p. 113 ff., and Schwaber/Sutherland 2017.
13 Further information on the comparison of known standards, e.g. in Kammerer et al. 2012, or projektmagazin 2014.
14 see Hüsselmann 2020, p. 6.

PMBoK Guide		ICB	
o	Body-of-knowledge approach	o	Body-of-knowledge approach
o	Emphasis on the behaviour of the PL	o	Emphasis on the (defined) competences of the PL
o	Defined knowledge areas	+	Inclusion of professional and
+	Process-oriented		contextual competences
-	No phase model	-	No concrete process models
-	No explicit artefacts	-	No concrete artefacts
...		...	

PM3		PRINCE2	
o	Body-of-knowledge approach	o	PM method approach
o	(only) exemplary artefacts	o	Formulated principles
-	Extremely comprehensive	+	Emphasis on phase orientation
+	Lots of material to work through		(but no phase model)
...		+	Concrete artefacts
		+	Business-case orientation
		+	Lessons Learned emphasis
		+	Distinct role model
		-	No fully comprehensive canon
		...	

DIN, ISO		Scrum	
o	PM method approach	o	PM method approach (process model)
+	Clear definitions	o	Operationalisation of agile principles
-	No consistency between DIN and ISO	+	Concrete artefacts
-	No concrete artefacts	-	Strong IT reference (cf. e.g. user stories)
...		-	No comprehensive PM concept
		...	
V-Modell XT			
o	PM method approach (process model)		
o	Operationalisation of waterfall-oriented principles (primarily)		
o	Extensive requirement for tailoring		
+	Concrete artefacts		
-	Strong IT reference		
...			

FIGURE 1.4 PM approaches in comparison (selected aspects).

Lean PM systematically and universally. The aim is to develop a universal PM framework with the following characteristics:

- Simple representation
- Easy applicability and concrete operationalisability
- Universal, project type-independent validity

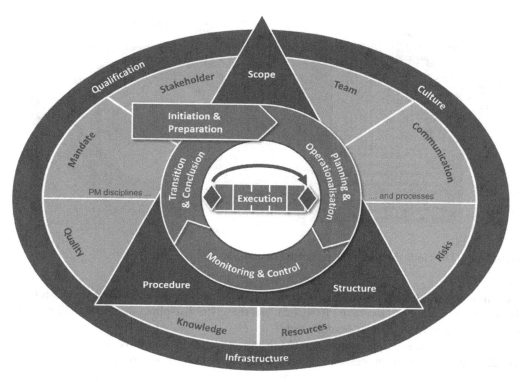

FIGURE 1.5   Big Picture of the Unified Project Management Framework.

- Best-in-class content, i.e. unification from existing PM standards
- Overcoming the (supposed) opposition of traditional and agile methods
- Possibility of adaptation to the respective project context.

The core of the Unified Project Management Framework is a process-oriented approach in which the activities of the PM are generally described in a recurring manner in the course of the project. These include ...

- first, the processes of *Initialisation & Preparation of* the project, with the central output of the project order, in the run-up to a project as well as
- recurrently during the course of the project, the processes of *Planning & Operationalisation* and
- accompanying the execution, the technical work of *Monitoring & Control* and finally
- *the Transition & Completion*, which focuses on the phase transitions within the progressive project processing as well as the transfer of operations.
- Complementary to the PM processes is the execution of the technically advancing project work, which directly serves the project purpose and is thus to be distinguished from the PM processes.

The focus for the above-mentioned processes and process groups is provided by the *PM disciplines*, which include, for example, the management of risks, stakeholders, project scope,

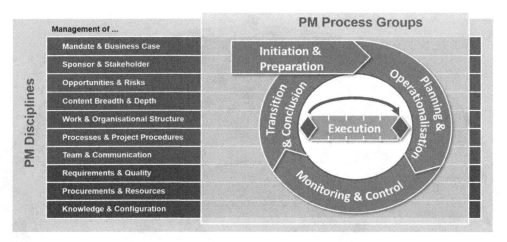

FIGURE 1.6 The Unified Project Management Framework.

etc. These PM disciplines are named and structured differently in the known PM standards (e.g. *knowledge areas in* PMI, *topics in* PRINCE2). In a sense, the disciplines described in Figure 1.6 can be seen to have a common denominator.[15]

Following a general system for classifying processes in the company, PM processes are divided into (1) strategic processes, (2) core processes and (3) enabler processes:

1. Managing the mission and the business case, the stakeholders including the client and the risks and opportunities.
2. Managing the project scope, project work and organisational structure, project workflow and processes as well as the team and communication.
3. Managing requirements and quality, financial and material resources in the operational, dispositive sense, as well as the configuration of project artefacts and knowledge preservation and utilisation.

In the design, this results in a matrix (disciplines x process groups) in which the activities are processed with varying intensity depending on the situation. It should be noted that the processes in Figure 1.6 can of course also be carried out in parallel and recurrently – not least in the sense of a Plan-Do-Check-Act-cycle (Deming circle).[16]

The description of the elements of the Unified Project Management Framework then includes the definitional version, the associated activities or sub-processes, typical methods and tools, inputs and outputs, process triggers, suppliers, performers and customers, as well as success factors and competences.

As shown in Figure 1.5, the Unified Project Management Framework must be embedded in the organisational framework for the company's projects. In addition to the infrastructural

---

15 after Hüsselmann et al. 2018, p. 13.
16 cf. Deming 1982, p. 88; Note: This corresponds to the process-oriented philosophy described in the PMBoK Guide, cf. PMI 2017, p. 50 ff.

requirements (material and immaterial resources such as premises, IT, licences, material, information and others), the qualifications of those involved in the project (PM techniques, professional, business context, behaviour/personality)[17] as well as those belonging to the cultural factors of the organisation (bureaucratic, agile, hierarchical, etc.) should be mentioned here, as these are crucial to include in terms of design, applicability and acceptance.

The following benefits can be associated with such a framework:

- Clear, easy to understand structure
- Framework for the systematic derivation of lean potentials and allocation of lean elements (principles, methods, tools)
- All important aspects together, e.g. process orientation, agile elements, knowledge areas, business orientation, etc. – by avoiding the weaknesses of individual standards
- Operational guideline and "toolbox" – without neglecting the superordinate overall picture
- Universally applicable and available – one framework for all projects in the company, e.g. without directional conflict "classic vs. agile".

1.2 Basics of Lean Management

Lean Management has its origins in the *Toyota Production System* (TPS) which was largely designed by the engineer and manager Taiichi Ohno. Womack and Jones took up the findings, expanded them through a multitude of other company analyses and developed the Lean Production and finally the general Lean Thinking approach. Due to the now diverse literature and other sources on the topic, only a slim overview of this is given below and the essential elements are presented that are important for conveying the idea and understanding the subsequent transfer to projects.

1.2.1 Origin from production

Lean Management is an extension of the Lean Production approach to the complementary processes of production. Lean production, in turn, is based on the Toyota Production System. Its roots go back to the 1940s, when Ohno began to change the production system at Toyota through his innovative ideas. Over the next decades, the Toyota Production System was successively expanded under his leadership to include various techniques that are now generally counted as part of the Lean Management concept. During this time, Toyota was able to rise to become the world's leading automobile company and has since served as an example for competitors in Asia, Europe and the USA.

The core idea of the Toyota Production System was and is the avoidance of *muda* – waste – of any kind. In the Toyota Production System the following techniques in particular were developed, applied and optimised, and are also among the core elements of Lean Management today:

---

17 cf. GPM 2017, p. 5 and p. 39 ff.

Jidoka	Automatic belt and machine standstill when a fault occurs
	Authorisation of staff to stop machines immediately if a fault occurs
	Separation of human and machine work, so that one employee can operate several machines or processes
Andon	Simple visual signal that draws attention to the status of a machine or process.
Kanban	Consumption-oriented just-in-time material provision with the help of a card system
	Self-controlling according to the pull-principle
Value stream orientation	Concentration on the production processes in the arrangement of the machines (instead of according to the type of machining)
Target costing	Backward-driven *design to cost,* in which the original equipment manufacturer (OEM) determines the value of a component for the end customer.
Kaizen	Process of continuous improvement in pursuit of perfection
5 Why	Repeatedly asking why to uncover causes of problems

Ohno's overriding goal was ultimately to deliver the required quality at minimum total cost and the shortest possible lead times. These are the aspects on which the Toyota Production System focuses. It should be emphasised that the domain of the Toyota Production System is characterised as series production, i.e. with frequently recurring, uniform processes of service provision. This is also the reason for the endeavour to standardise processes at an ever higher level of performance and stability in the Toyota Production System and the Lean Management approach. This is a particular difference to the domain of PM, which must be taken into account when transferring ideas and methods. By definition, projects are characterised by uniqueness, or at least not a uniform repetition of the same old thing. Appropriate attention is paid to this aspect in the design of Lean PM (see Chapter 6). But first, back to the roots of Lean Management.

Womack and Jones integrated the Toyota Production System into their study of "lean" production in the automotive industry, which was carried out worldwide in the 1980s.[18] With the help of the benchmarks available, they later derived the general principles from this study, which they published as Lean Thinking elements for the first time in 1996.[19] The idea was to develop an approach from the findings of successful production management (especially at Toyota) that would provide managers with an overarching but also pragmatic guideline on how to apply the many different individual techniques. In particular, they formulated the five core principles of Lean Management, which we refer to in the rest of the book. Furthermore, a number of additional lean elements were described, which have been added to over the years by science and practice. This is in the nature of the concept, as we will see later on, and in a way creates a never-ending story of the development of Lean Management.

---

18 see Womack et al. 1991.
19 see Womack/Jones 2013.

Part of this story is the transfer of Lean Management to other areas of the company. Womack and Jones had already started with this by defining the lean enterprise, in which Lean Management is implemented beyond production and eventually beyond the company's boundaries in the entire value chain, especially the logistical supply chain. Subsequently, the Lean idea was explicitly transferred to other industries and further corporate functions, which from the production point of view are referred to as indirect areas: Lean Procurement, Lean Sales, Lean Administration, Lean Innovation as well as Lean Construction, Lean (Software) Development and also Lean Start-up. Particularly noteworthy here is the construction industry, where the construction crises in the USA in the 1980s led to the development of the first transfers to project management – *Lean Project Delivery*, which has been developed in a leading role by Glenn Ballard at the Lean Construction Institute of the University of California since around 2000.[20] In the field of software development, the second major source of inspiration for the development of the present Lean PM concept, the Poppendiecks, who also presented their *Lean Software Development* approach in the USA at the same time, deserve special mention.[21] They call the approach *agile toolkit,* which illustrates the closeness of the concepts of agile management to Lean Management. I will discuss the differences in my opinion in more detail later on.

But let's first take a brief systematic look at the central elements of Lean Management.

1.2.2 Identification of waste

The primary postulate of Lean Management is the avoidance of waste (*muda*) of any kind in the company.[22] Waste is generally defined as all activities and processes that generate costs but no value (for the customer). Basically, the following types of waste are to be distinguished:[23]

Process-related waste:
This includes activities that do not directly create value for the process customer but are (currently) necessary for the execution of the service creation process. Examples are planning activities or set-up processes. The aim is to minimise these to what is really needed, up to and including possible complete elimination.

Business waste:
These are (secondary) processes that do not directly add value for the customer but are necessary for the business as a whole. Examples are financing processes or personnel administration. These need to be reviewed in terms of their efficiency and, if applicable, their elimination.

Pure waste:
This refers neither to processes that directly create value nor to those that support or enable value creation. Examples are an unnecessary flood of paper or expenses. These are to be eliminated immediately.

In the field of Lean Production, as the historical basis of the Lean Management approach, Taiichi Ohno already identified the seven causes of waste known by the catchy acronym *Tim*

---

20 see Ballard/Howell 2003 (first known essay).
21 see Poppendieck/Poppendieck 2003.
22 cf. Womack/Jones 2013, p. 23.
23 cf. Bicheno 1998, p. 9.

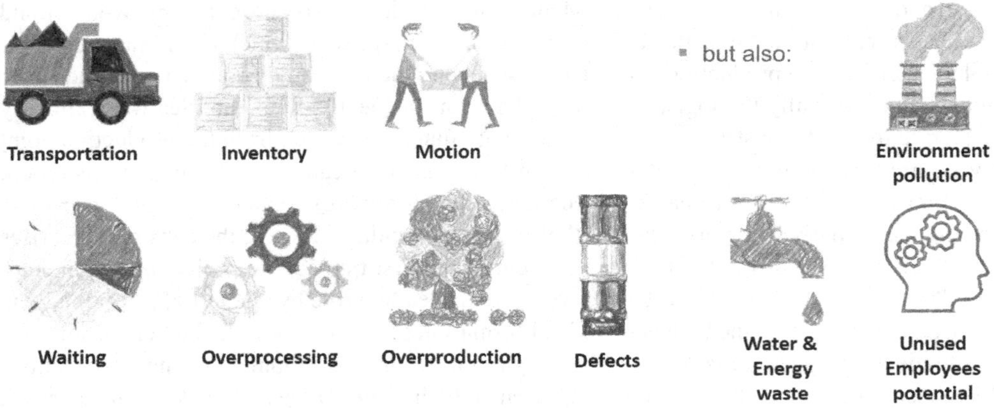

| Transportation | Inventory | Motion | | but also: | Environment pollution |

| Waiting | Overprocessing | Overproduction | Defects | Water & Energy waste | Unused Employees potential |

FIGURE 1.7 Waste in the production process Tim Wood +.

Wood (transport, inventory, motion, waiting, overprocessing, overproduction, defects) (see Figure. 1.7).[24]

Other causes of waste have been identified further on. These include in particular: Environmental pollution, unused employee potential, and water and energy waste.[25]

However, redundancies or repetitions, which are generally regarded as waste in Lean Management, are sometimes quite useful in complex projects, because they increase the certainty of the result (e.g. test runs). In this respect, maximum efficiency, such as the avoidance of any redundancy, is therefore not conducive to achieving the goal, which makes it difficult to draw appropriate boundaries in the application of Lean principles. Redundancy can reduce risks and avoid accidents. A report on the development of Lockheed Martin's F-22 fighter makes it clear that a new conception of value and waste is called for, as activities added to a process can serve to catch problems, before they are cascaded through many other activities, or increase confidence in the desired outcome. Then they are value-adding, despite their characterisation as non-value-adding according to the traditional definition of Lean.[26]

1.2.3 Core principles of Lean Management

In the diverse literature on Lean Thinking and Lean Management, a number of terms and slightly different structural designations have developed since the first publication by Womack and Jones. For example, the number of terms for the so-called *Lean Principles* varies between 5 and 14.[27] Methods and principles are linguistically mixed, etc. Therefore, the understanding of the underlying terms is first fixed in the following.

---

24 see Ohno 2013, p. 163.
25 see e.g. Bertagnolli 2018, p. 26 ff.
26 see Belvedere et al. 2019, pp. 411–413.
27 cf. Liker 2007 and Womack/Jones 2013, p. 23 ff.

FIGURE 1.8   The five core principles and the central paradigm of Lean Thinking.

DEFINITION OF LEAN MANAGEMENT

Lean management is the consistent orientation of operational processes towards the customer by reducing them to what is of value to the customer.[28]

In this sense, ***avoiding waste*** is the central ***paradigm of*** Lean Thinking (see Figure 1.8). Womack and Jones formulate five basic principles of Lean Thinking in the original: *Value creation, value stream orientation, flow, pull* and (striving for) *perfection.*

The following explanations for the clarification of terms essentially quote the corresponding definitions and selected explanations by Womack and Jones as the originators of the concept.[29] We deliberately refrain from providing our own explanation here, as these are used as original sources in the development of the Lean PM concept.

Value: *A service delivered to the customer on time and at an acceptable price, defined in each case by the customer.* The crucial starting point of Lean Thinking is value. Value can only be defined from the perspective of the end user. And it makes more than sense for it to be defined in terms of a specific product (product or service, often both at the same time) that satisfies the customer's need at a specific price. The value is created by the manufacturer.

---

28 Following Gorecki/Pautsch 2013, p. 8.
29 Womack/Jones 2013, Glossary, p. 404 ff. and selected passages on this subject.

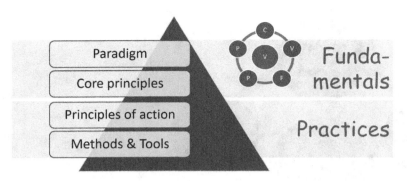

FIGURE 1.9 Hierarchy of Lean elements.

Value stream: *The specific activity required to design, order and deliver a particular product, from concept to launch, from order to delivery, and from raw material into the hands of the customer.* The value stream consists of all the specific activities required to take a given product through the three critical management tasks in any organisation: Product development, information management, and the physical transformation from raw material to a finished product in the hands of the customer.

Flow: *The progressive achievement of tasks along the value stream so that a product progresses from design to launch, from order to delivery, and from raw material to the hands of the customer without interruption, scrap or returns.* The idea of flow is to define the work of functions, departments and companies in such a way that they can make a positive contribution to value creation, taking into account the real needs of the product and of the employees at each point along the flow, so that the value stream flows continuously.

Pull: *Cascading production and delivery instructions from downstream to upstream, according to which nothing is produced at an upstream location until the downstream customer registers a demand.* The opposite of push, pull is the ability to design, plan and manufacture exactly what the customer wants and when they want it. This means that companies only produce what the customer demands. In simple terms, pull means that no one should produce a good or provide a service at an upstream stage until the downstream customer demands it.

Perfection: *The complete elimination of waste (muda) so that all activities along the value stream generate value.* Womack and Jones refer to the aforementioned principles as the original principles, which stimulate each other in a circle and strive for perfection in a seemingly endless process. *Kaizen* calls for a culture and a process of continuous improvement.

In addition to the above-mentioned five core principles, which we refer to in the following as key **design principles,** authors use a variety of other methods and tools, but also so-called principles in the design of Lean Management.[30] In the course of a systematic derivation of the Lean PM approach, we distinguish conceptually between the levels of *principles of action*, *methods* and *tools* (see Figure 1.9).[31]

---

30 see e.g. Bicheno 1998 or Herbig 2015.
31 see Hüsselmann et al. 2018, p. 10.

These are defined as follows:

Following the *Capability Maturity Model Integration* (CMMI),[32] **principles of action** are defined as *practices* or *best practices* that serve in a proven way to implement the design principles, i.e. to achieve the corresponding goals. In general, practices serve generically to support the fulfilment of all principles and are not dedicated to a single principle.[33] Examples of principles of action are ...[34]

- *Gemba* (Japanese for "place of action") expresses that problems are analysed, processed and solved where they occur. This is the best place to analyse, evaluate and optimise processes.
- *Shitsuke* (Japanese for "Keeping all points and constantly improving") is applied within the framework of the *5S/5A* to improve the workplace environment. *5S/5A* is a grouping of terms through which waste is avoided at the level of the individual workplace through standardisation of procedures and operations (see also Section 4.7.5).

Methods serve to give the user concrete recipes in the operational implementation of the principles and practices, as it were, which can be applied operationally in optimisation projects and workshops. An example of a method is *value stream mapping* (VSM) or *target costing*.[35]

Finally, **tools** are operational aids that enable the use of methods. For example, *Makigami* is a possible tool for conducting value stream mapping in a Lean Administration context.[36]

---

32 CMMI is a family of reference models for different application areas such as software or product development or service delivery.
33 cf. Herneck/Kneuper 2011, p. 15.
34 according to Herbig 2015, p. 26, p. 105 and Gorecki/Pautsch 2013, p. 73.
35 cf. e.g. Bicheno 1998. Value stream mapping is used, for example, to visualise value-adding and non-value-adding activities in the service production process.
36 see Sunday 2015.

2
AGILITY AND COMPLEXITY

After reading this chapter, you will know …

- the definition, origin and core principles of agility,
- the characteristics of "hybrid" approaches within projects
- a brief introduction and classification of Scrum as the most popular method, and
- the definition and challenges of, and approach to handling complexity.

2.1 The essence of agility

2.1.1 Motivation

In the context of organisational, process and project design, agility – along with digitalisation – currently appears to be the buzzword par excellence. More than a million hits on Google search speak for themselves. Everything is supposed to be or become agile, and all the elements of agile approaches, especially Scrum in particular, are referred to as agile – sometimes unreflectively. But what is really behind it? Which elements of so-called *agile methods* are really agile and what constitutes agility at its core? Under the keyword *hybrid* such agile elements are increasingly mixed with so-called *classic* ones in order to realise a "best of" – at least for the context of the organisation under consideration. In the context of the systematic development of a Lean Project Management approach, which sees itself as a further evolutionary development of the PM discipline, it seems to make sense to take a substantial look at the terms.

DOI: 10.4324/9781003435402-3

2.1.2   Origin of the term

Today's use of the term *agile* or *agility* can be traced back to various roots – from its etymological origin to its use in management theory to the birth of its current prevalence in the context of the design of modern software development processes. In purely linguistic terms, the Duden, for example, explains the term *agile* with the synonyms nimble, nimble and flexible. This obviously is aligned with the use of the term in management and IT development in recent decades.

Korn gives an overview in his essay *"The 'agile' approach: New wine in old bottles – or a 'déjà vu'?"* (in German) of the various origins and interpretations of the term in the above-mentioned context.[1] A central element is the interpretation of agility as flexibility and adaptivity by means of empirical process control. This is in contrast to process control in the sense of an extensive, pre-planned process definition with subsequent control (*command & control*). Building on this, Korn identifies definitional enrichments of the term with principles of Lean Management and self-organised collaboration.

He also identifies another, more systemic definition of the term in military terms. For example, in connection with the (agile) leadership of social systems, the US Department of Defense demands *robustness, resilience, responsiveness, flexibility, innovativeness* and, last but not least, *adaptability in* the execution of high-risk and complex missions. The acronym *AGIL* according to Parsons, which was already defined in the 1950s, also falls into this category: *Adaptation to the environment + Goal Attainment + Integration of the subsystems for performance + Latency/ Structural Maintenance to preserve the identity of the system.*

The most prominent source of current discussions about so-called *agile PM* seems to be the *Agile Manifesto* from 2001.[2] In this manifesto, a number of experienced software engineers drew on their wealth of experience to formulate four central values of agile software development and twelve corresponding principles of action. In the sense of the above-mentioned cascading definition of agility, not all of these values and principles can be identified as genuinely agile. However, the principles mentioned above clearly include the value "We value reacting to change more than following a plan." Contrary to many interpretations, the authors are not saying, for example, that there is no longer a plan to be created: "Although we find the values on the right important, we value the values on the left more highly."

Genuinely agile principles of action can also be found, such as: Changes in requirements even late in development are welcome. Others can be clearly identified as a means to an end, such as: Constant attention to technical excellence and good design promote agility.

Finally, agile thinking led to derivative approaches. The most prominent of these (cf. e.g. study *Status Quo Agile*)[3] is Scrum, which was developed in its current form by Schwaber and Sutherland and first documented as early as 1995. Nevertheless, it should first be noted that in the authors' *Scrum Guide* the word *agile* occurs only once, namely in the identification of the Scrum Master's task to convey the correct understanding of agility. The definition of Scrum as a framework for the development, delivery and maintenance of complex products, within which people can tackle complex adaptive tasks, is particularly revealing. Last but not least, Schwaber

---

1 see Korn 2013.
2 see Beck et al. 2001.
3 see Komus 2015.

and Sutherland thus refer to Scrum as a process framework for managing work on complex products – and not as a PM method. Rather, the complexity of the products and adaptivity in task processing are obviously the core features of Scrum.[4]

2.1.3 Agility – a definition of the term

With the above in mind, an attempt can now be made to get to the bottom of agility ... or rather, let us say, to structure the term and the associated principles, practices and methods a little better.

Flexibility and adaptivity as paradigms of agile action
In essence, agility can be defined as the ability of an operational organisation (project, company, division) to adapt flexibly and quickly to changing influencing factors over the course of its processes – without losing sight of the overall objective. Changing influencing factors can be caused by changing environmental conditions (e.g. the market), changing stakeholder-requirements (e.g. of the users) or knowledge gained in the course of processing (e.g. with regard to the procedure). The aim of dynamic adaptation is to optimise the benefits, efficiency and effectiveness of action in the course of the process.

In agile, change is seen as immanent, so that the possibility of adapting the system is anchored as a fundamental component. This anchoring has both a methodological-procedural (procedures) and a mental (personal) and sociological (interaction) foundation. In contrast, non-explicitly agile systems act more statically and do not regard changes as an immanent component, but rather as disturbances. Their handling in these systems, although systematically enabled, tends to be operationalised as an exceptional procedure (cf. Change Request Management).

The temporal stability of the system is basically considered desirable – as Peter Kruse says: "A system in which everything is constantly flowing is psychotic!" – as this generates productivity and security, especially in the sociological, organisational context (e.g. planning security). Agility is therefore needed when dynamic changes in the influencing factors conflict with the fundamental desire for temporal stability of the system. The so-called *classical* project approaches can (and should) rather be characterised as *plan-driven*. Planning is not unagile as long as it follows the *JEDUF principle* (*just enough design up-front*), i.e. it is rolling and remains adaptive.

2.1.4 Core principles and practices of agility

With the aforementioned considerations, flexibility and adaptivity are seen as genuine and thus as central paradigms of agility. These can be complemented by principles of action and organisation that have proven to be fundamental and critical to the success of achieving agility. Figure 2.1 illustrates this relationship schematically.
As generic core principles for operationalising agility, therefore, ...

1. the extensive **cooperation** between and with all stakeholders of the measure, especially in the team,
2. the extensive **delegation** of responsibility, to the level of maximum professional competence, up to the self-organisation of the teams where possible,

4 see Schwaber/Sutherland 2017.

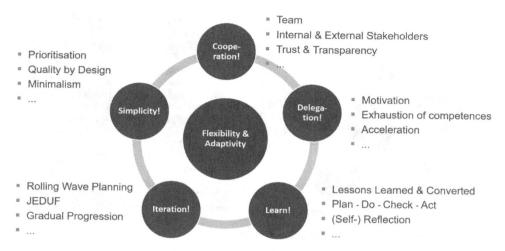

FIGURE 2.1   The core principles and the central paradigm of agility.

3. regular and short-term reflection on results, processes and developments to enable CIP **learning**,
4. rolling **planning** and design, i.e. as much design and planning in advance as necessary – and as little as possible, and
5. the demand for **simplicity** in the sense of qualitatively robust solutions that meet the requirements of customers can be identified.

The implementation of these core principles thus consistently leads to the achievement of flexibility and adaptivity, i.e. agility. However, they are not genuinely part of agility, because who would seriously dispute that points 1 to 5 are not also useful, even obligatory, to be applied in the so-called classic PM standards according to PMI, IPMA or PRINCE2 (even though they are often not found in practice). In the operationalisation of these principles of action and organisation, a multitude of practices have emerged, as they have been manifested, for example and not least, in approaches such as Scrum. These include, for example, the *Minimum Viable Product*, the *work-in-progress limitation*, the *daily standup*, the *task board, user stories*, the *burndown chart, time boxing* or *design-to-budget*.

In the course of the development of Lean PM as the methodological superstructure of modern PM, it is easy to find a multitude of points of contact in the explanations given above – but also differences: Lean is not identical with agile! For example, Lean Management postulates the central paradigm of *avoiding waste*. Core principles can also be identified that contribute to the implementation of this paradigm. These have already been described by Womack and Jones and are: *Value from the customer's point of view, identify value stream, implement the flow principle and the pull principle* and finally strive for *perfection* (see Section 1.2.3). The differences to the preconceived ideas of systematising agility are obvious. Nevertheless, there are overlaps and above all potential with regard to the application of practices – the Kanban board (see Section 4.5.2) is the best example of this.

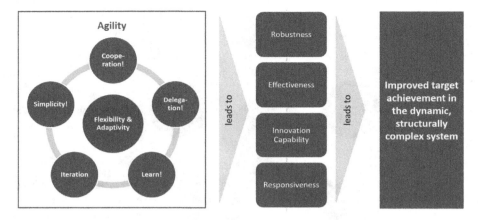

FIGURE 2.2 Agility – paradigms, principles and goals.

Flexibility and adaptivity by means of empirical process control can be identified as the characteristic substance of agility. Derived from this, a number of core principles emerge in the operationalisation of agile approaches, as well as a multitude of possible suitable practices in their implementation.

From a more strategic point of view, robustness, resilience, responsiveness and innovation are characteristics of agile systems at the system level, which ultimately serve to achieve the higher-level goals of the system (company, project, etc.) (see Figure 2.2).

In this context, the antithesis to empirical process control – *planning, command & control* – is not a contradiction, but rather a context-related option or even a necessity (see Section 4.8). The (temporal and structural) complexity of the measure or its organisation and control is a central indicator for the use of planning, command & control.

Lean PM pursues the approach of generalising the above-mentioned perspectives and providing project (management) designers with a toolbox that enables them to implement situationally tailored approaches to efficient and effective project handling – without having to fight the trench warfare between "new and old" PM.

2.1.5 Hybrid approaches

Based on the previous considerations, the following two poles can be identified with regard to PM approaches:

plan-driven – with an emphasis on stability and planning security – as well as agile – with an emphasis on flexibility and adaptability.

In contrast, the identified principles of action and organisation as well as the practices cannot be clearly assigned, even though they naturally promote corresponding objectives or, viewed the other way round, make them more difficult to achieve if missed. As examples, two of the basic principles according to PRINCE2: Ongoing business justification and learning from experience – are also undisputed elements of agile approaches.

Various studies and practical experience show that a hybrid mix of methods is often used depending on the company and context, i.e. standardised procedures are not applied in their originally conceived pure form.[5] Many authors and studies speak of so-called *hybrid* approaches. Therefore, an attempt will be made here to define the term.

Komus and Kuberg describe the hybrid approach in their study "Status Quo Agile" as a mixture of agile and classic PM methods, interpreted in the sense of moderate agile methods.[6] Kurtz and Sauer describe hybrid PM methods somewhat more specifically as those that combine agile and classic techniques.[7] In contrast, Sandhaus et al. focus in their book on hybrid software development on the sequential combination of classic phase models (e.g. Waterfall) with agile methods, such as those used in Scrum.[8]

Following Timinger's definition and taking into account the previous considerations, the following definitional description emerges:[9]

HYBRID PROJECT MANAGEMENT

Hybrid project management is the use of practices (processes, methods, tools), organisational forms (structures, roles with their tasks, powers and responsibilities) and phase sequences of different standards or procedure models in a project. This is done with the aim of taking into account the contextual conditions (degree of complexity of the measure, capabilities of the organisation, formal framework conditions) in the best possible way to achieve the project goals.

In particular, plan-driven and agile elements are often mixed. In their empirical study on process models and methods in hybrid PM, Blust and Kan identify 113 (!) hybrid process models that are used in practice by the companies surveyed.[10] The combination of agile and classic methods – especially in product development and organisational projects – is in first place. Most users (approx. 55%) rate the hybrid model as successful, only about 15% state that the hybrid model does not work well. The hopes that users associate with the use of hybrid models are very heterogeneous and of course depend on the combination chosen in each case. As examples, the satisfaction of a heterogeneous group of stakeholders (combination of Scrum and critical chain management) or the promotion of agility and self-organisation in a stage-gate process required by regulation (combination of Scrum and stage-gate process).

Overall, various forms of hybrid PM approaches can be applied and observed in practice. Particularly with regard to the combination of classic and plan-driven approaches, these are the following modus operandi:

---

5 see Blust 2019.
6 see Komus 2017.
7 see Kurtz/Sauer 2018.
8 see Sandhaus et al. 2014.
9 cf. Timinger 2015, p. 241 ff.
10 see Blust/Kan 2019.

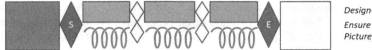

Design-to-Budget
Ensure conformity to the Big
Picture of the project

FIGURE 2.3 Bimodal procedure model.

Gaining planning and solution
certainty

FIGURE 2.4 Evolutionary process model.

Cherry-picking

FIGURE 2.5 Best-of-breed approach.

Bimodal: Here, individual sub-projects with different procedures are carried out in correspondingly extensive projects (see Figure 2.3). For example, in an SAP implementation project, standard parts can be handled in waterfall mode (business blueprint → customising → data transfer, etc.), while additional developments – in line with the original reason for agile software development – are carried out according to Scrum development.[11]

Evolutionary: The mode of the project changes in the course of the project (see Figure 2.4). For example, due to uncertainties at the beginning of the project, a few sprints are initially carried out until sufficient certainty has been achieved with regard to the procedure and solution, i.e. the complexity has been reduced.

Best-of-breed: Solution elements from the different approaches are synthesised into a new method, optimised according to the circumstances – e.g. timeboxing within the waterfall approach (see Figure 2.5).

Combinations of the above-mentioned characteristics as well as further derivations are conceivable. The ability of an organisation to live with different "operating systems" at the same time is also referred to as *organisational ambidexterity*. Ultimately, the question arises as to when it makes sense to use which combination. The answer can be found not least in the explanations on the project-specific *adaptation of the PM system* (see Chapter 6) and the *Agilometer* that has been developed (see Section 4.8). Overall, it can be said that

Lean PM is adaptive hybrid PM with guidelines!

This means that the systematic or situational application of Lean Management (values and practices) results in a purposefully designed PM – generally in the form of a hybrid system.

11 cf. Hüsselmann 2014.

In the following, some examples of hybrid design are presented, which can be classified as structural, temporal and contextual ambidexterity.

2.1.6 Examples of hybrid project approaches

2.1.6.1 Bimodal design

The bimodal design of a PM system means that several types of management are practised simultaneously within the project. In large projects with several sub-projects, this typically concerns the different approaches in the various sub-projects (so-called *structural ambidexterity*).[12] Kirchhof and Kraft report on an application in an IT project in which a host-based legacy application was to be made accessible via a modern web front-end.[13] The project was structured organisationally according to technical aspects into the sub-projects *SP 1 Frontend, SP 2 Security, SP 3 Hostconnect* and the *Test Centre*. Sub-project 1 was handled according to Scrum while sub-project 2 and sub-project 3 as well as the overall project were structured according to the classic software development process model.

This meant that different processes and roles were used in the areas of work assignment, reporting, meeting structures, requirements and quality management, which had to be synchronised. In summary, the following forms of synchronisation can be identified:

Common Heartbeat:	The timing of phases and meetings was in weekly or four-weekly cycles. In this way, *jours fixes*, steering committee meetings and sprint procedures could be aligned with each other.
Integrative work:	People with key roles were involved in each other's sub-projects, requirements managers and testers were integrated early and across all sub-projects.
Results-oriented reporting:	A distinction was made between sub-project-internal and cross-project reporting. The former could be designed flexibly, e.g. with story point-based burndown charts, the latter was harmonised on the basis of work package (completions).
Common language:	In order to make the cooperation successful, language barriers had to be overcome (cf. UPMF, Section 1.1.2) and trust had to be created.

2.1.6.2 Evolutionary design

In the evolutionary approach, the procedure develops in the course of the project. For example, the so-called Water-Scrum-Fall approach is based on a model with phases that build on each other. In the meantime, iterative-incremental sections are integrated into this model according to Scrum.[14]

An application example is described by Fahr et al. for the SAP S/4HANA implementation model of BearingPoint, a Europe-wide acting management and technology consultancy.[15] SAP

---

12 cf. Hüsselmann 2014.
13 see Kirchhof/Kraft 2020.
14 see e.g. Komus 2020, p. 84.
15 see Fahr et al. 2019.

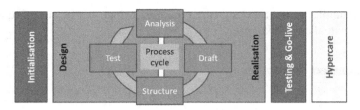

FIGURE 2.6   Hybrid five-phase approach to SAP S/4HANA implementation.

S/4HANA is the fourth generation of SAP's ERP system, based on SAP's in-memory database technology.[16] BearingPoint reports on a number of SAP S/4HANA implementation or migration projects in which a hybrid approach, as shown in Figure 2.6, was applied. This involves alternating between a classic, sequential phase model and an iterative approach.

In Figure 2.6 the sequential, basically waterfall-like, phases of *initialisation* and *testing & go-live* are shown in dark grey, and the agile phases, carried out in Scrum-methodology are shown in medium grey, supplemented on the right by an initial operating phase (hypercare).

The *initialisation* phase basically contains the activities that, according to the Unified Project Management Framework, are included in the process group Planning & Operationalisation (and not in the process group Initiation & Preparation). This includes clearly defined project guidelines, team formation including training, a confirmed project plan, a defined communication and documentation plan as well as a high-level risk assessment.

In the *design* phase, acceptance criteria are defined and a product backlog is created with prioritisation of the user stories. The starting point is the *Business Process Master List* (BPML), which contains the documented processes as well as a technical specification (which may come from a preliminary project). The user stories are described by comparing the requirements with SAP S/4HANA processes and enriching the requirements with fulfilment criteria for later acceptance. The user stories are prioritised and included in the backlog. In the process, the extension of the required standard processes is checked due to possible additional requirements, the necessary architectural additions and supplements to the business concepts are developed and documented for the interfaces, and finally a migration concept is developed. Each user story is assigned to different work packages for implementation, e.g. the creation of key user documentation or the realisation of customising. The implementation of user stories is summarised in releases. Depending on the scope of work in the implementation, the number of user stories per release varies.

Based on the product backlog the SAP S/4HANA system is realised in various sprints on the basis of the user stories, which are bundled into work packages. Within the framework of sprint-planning, the work packages are selected for the sprint along with the conception of how the work packages are to be implemented. This is followed by the *realisation* of the work packages defined in the planning, including documentation, testing and monitoring of the implementation progress.

After every sprint, the requirements are tested against the acceptance criteria and regression tests are carried out. Depending on the scope of the sprint, sprints may also be combined for releases. The "Go-live" of user stories is summarised in releases. Finally, the efficiency and

16 ERP: Enterprise Resource Planning.

effectiveness of the working method during the sprint is checked and, if applicable, agreements are made on an optimised procedure for the next sprint.

The goal of the **Testing & Go-live** phase after completion of the implementation sprints is to prepare for the Go-live of the SAP S/4HANA system. In this context, comprehensive end-to-end integration tests are carried out across all modules, followed by knowledge transfer – first via the key users and then to the entire local organisation. For this purpose, the end users are trained on the basis of training documents. Furthermore, the data migration and the cut-over phase are prepared and carried out, which requires meticulous planning.

The final **Hypercare phase** is already part of the system operation, although final project activities are carried out. These include assisting with the start-up of the end users' daily routines and the on-site support organisation (Gemba), troubleshooting as part of bug fixing, and conducting formal acceptance.

According to BearingPoint, the hybrid approach to SAP S/4HANA implementations described above enables it to realise the following benefits:

Speed:	Available result after each sprint
	No waiting for certain project phases, e.g. to test with all modules
	Measurable increase in the effectiveness of the implementation team
Cooperation:	Very close cooperation of the team to implement requirements
	Continuous involvement of the department in the implementation process
	Joint learning and continuous improvement of the approach
Step-by-step construction:	Iterative approach in sprints (1-month cycle)
	Go-live requirements are tested according the acceptance criteria and released by the team

2.1.6.3 Best-of-breed design

As the name suggests, this variant involves cherry-picking the practices that appear to be the most effective in the respective context. Lean thinking should be the guiding principle for selecting practices according to the problems to be solved.

Examples for this best-of-breed design of the project approach are provided in Chapter 5 from my own project history. In this respect, I will refrain from repeating what has been described there and instead refer to the reading of this chapter. However, due to its outstanding significance with regard to the success of the project, which was evaluated at the end, the very innovative timeboxing should be mentioned once again. Timeboxing was used within the classically carried out project to clock in over half a year on a week-by-week basis. A repetitive cycle of five *sprints* with defined content was carried out, structured according to the requirements of the technical and organisational integration of the roll-out of services. A particular cycle was related to an associated process area, e.g. payroll accounting – which represented the process-oriented product increments that were transferred to the productive system at the end of each cycle. When working on the timeboxes, a backlog previously set up by the technical project management was followed. If an element of this backlog (to-do) remained unfinished within the timebox, the technical project management decided whether the item had to or could be removed, provisionally clarified in an analogy conclusion or reworked in the following cycle. The flexibility thus created – even in the case of a contract-for-work constellation (see Section 5.1) – led to increased satisfaction on the part of the stakeholders and a high degree of (temporal) planning security for those involved.

2.2 How much PM is in Scrum?

In addition to the context-related application of hybrid models, i.e. based on the question of which procedure is best suited in a specific case (see also Chapter 6), as mentioned above, a number of agile models have gaps with regard to comprehensive PM that need to be filled in any case. Scrum is used as an example to illustrate this.

2.2.1 The Scrum Framework

Scrum is currently the most widespread working method in the field of agile project approaches.[17] In many cases, Scrum is referred to as the PM approach, i.e. the method by which PM is carried out in companies. Scrum practices are also increasingly being applied to project types outside of software development projects.[18]

The first thing to note is that Scrum is not a complete PM approach and the authors Schwaber and Sutherland (Scrum Guide) do not even pretend that it is. Rather, Scrum was developed by them as a "framework for developing, delivering and sustaining complex products". "Scrum is neither a process, nor a technique, nor a complete methodology."[19] Scrum is a process practice that can succeed in operationalising the values of the Agile Manifesto. It should be noted that Scrum essentially describes the organisation (process and structure) of software development. For example, the concrete practices of software engineering etc. are also missing. The slimness of the original publication – it is only 20 pages – by Schwaber and Sutherland on Scrum, the Scrum Guide, ultimately testifies to this ... and also constitutes a strength of the approach. This strength ultimately leads to the ideas of Scrum being transferred to other types of projects, or at least to an attempt to do so.

At its core, Scrum according to the Scrum Guide consists of a set of predefined procedures (events), roles and artefacts. These are as follows:

Roles:	(Development) Team	Product Owner
	Scrum Master	
Procedures:	Daily Scrum (Daily Stand-up Meeting)	Sprint Review
	Sprint Planning	Refinement*
	Sprint Retrospective	
Artefacts:	Product Backlog	Impediment*
	Sprint Backlog	User Story*
	Increment Burndown-/Burnup-Diagram*	Release*

The entries marked with * are not defined by Schwaber and Sutherland in the original as a formal element of Scrum (role, event or artefact), but are mentioned there or have become established as common practice. In addition, the principle of timeboxing, i.e. the strict alignment of the

---

17 see Komus et al. 2020.
18 see e.g. Feldmüller/Sticherling 2016.
19 Schwaber/Sutherland 2017, p. 4.

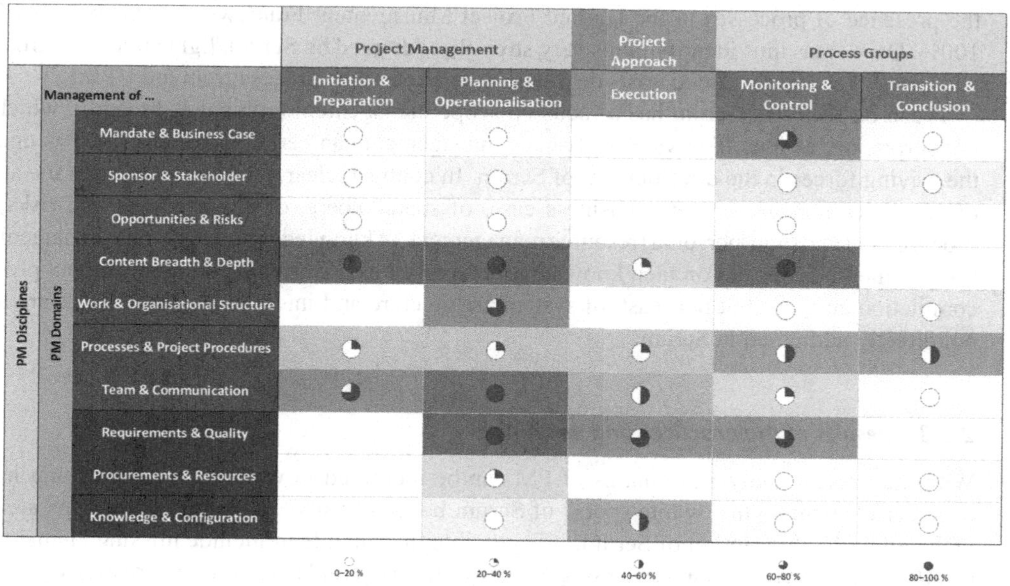

Management of ...	Project Management		Project Approach	Process Groups	
	Initiation & Preparation	Planning & Operationalisation	Execution	Monitoring & Control	Transition & Conclusion
Mandate & Business Case	○	○		◕	○
Sponsor & Stakeholder	○	○		○	○
Opportunities & Risks	○	○		○	
Content Breadth & Depth	●	●	◕	●	
Work & Organisational Structure		◕		○	○
Processes & Project Procedures	◔	◕	◕	◑	◑
Team & Communication	◕	●	◑	◔	○
Requirements & Quality		●	◕	◕	
Procurements & Resources		◔		○	○
Knowledge & Configuration		○	◑	○	○

Legend: ○ 0–20 %   ◔ 20–40 %   ◑ 40–60 %   ◕ 60–80 %   ● 80–100 %

(Row axis labels: PM Disciplines / PM Domains)

FIGURE 2.7   Mapping UPMF-Scrum, overview.

procedure with previously defined time windows, plays a major role in Scrum. This is expressed in particular in the sprint system, but also in the defined durations of all other procedures. In this way, a Pareto optimum can be achieved in terms of the output-effort ratio for meetings, etc.

2.2.2   The Delta Analysis

Since Scrum does not describe a complete PM method, the question inevitably arises as to where the gaps in Scrum actually lie and how these can be filled if necessary. To make a delta analysis, the Unified Project Management Framework (UPMF) is used (see Section 1.1.2).[20] This framework is designed as a process-related reference model for holistic PM and, due to its simple structure, offers the possibility to examine the degree of coverage of Scrum through a corresponding mapping. For each PM process defined by the UPMF it is identified whether it is or is not addressed by Scrum (0), whether elements are implicitly (1) or explicitly (2) included or whether Scrum even addresses the considered PM aspect in a particularly pronounced and targeted way, i.e. whether there is a focus or strength of Scrum here (3).

The following overview of results (see Figure 2.7) is based on the process analysis presented in Section 2.2.4 (see Figure 2.8). The fields of the UPMF matrix were weighted according to the process assignment (see process identifier).

It is possible to identify in which process areas Scrum makes direct contributions and where it does not. The grey gradation in the fields of the matrix also relativises the degree of support according to the scale described above. The maximum possible degree of support, measured by

20 Hüsselmann 2020.

the presence of processes in the Unified Project Management Framework, was normalised to 100%. Dark areas thus identify fields very strongly addressed by Scrum, light areas less strongly addressed. The lightly dotted areas do not contain a reference process from the UPMF.[21]

It can be seen that Scrum has a focus on scope management (content & breadth), which is not surprising, as the effectiveness of scope management can certainly be described as one of the driving forces in the development of Scrum. In contrast, clear gaps are revealed in the areas of (explicit) management of a business case, of stakeholders, of opportunities and risks, of procurement and (non-personal) resource management, of knowledge and document management ("Agile methods depend on tacit knowledge")[22] as well as of formal and organisational project completion and, last but not least, of system architecture and integration. These fields are also not directly addressed in Scrum.

2.2.3 Results of the practices and methods

With the present analysis, domains of PM can be identified in which Scrum can bring in no or fewer elements – the "white spots" of Scrum become visible. Although these gaps are not included in the description of Scrum, it is possible in principle to include missing elements as backlog entries in the actual processing of a project. An example of this is the creation of system documentation. If this is required by the project object, e.g. in a validated environment, then such a deliverable can also be created via a corresponding user story. In this way, white spots can basically be handled through the concrete design of the backlog. The scaled agile method *SAFe (Scaled Agile Framework)* also speaks here in a broad sense of so-called *enablers,* or structurally closes gaps, such as in architecture management.[23] However, it remains the case that the corresponding elements are not defined in the Scrum method and are therefore not predestined.

The grey and especially dark grey fields in the UPMF-matrix identify the areas with high coverage by Scrum elements. Due to the type of projects Scrum was developed for (development, delivery and maintenance of complex products in relatively small teams), this does not mean that an adequate practice is available with the corresponding Scrum elements in every case. For example, Scrum emphasises communication within the team and with the subsequent users of the products (customers), but does not provide direct guidance on communication with other stakeholders, e.g. political interest groups. In this sense, there is no complete communication management, even if the communication of the parties directly involved (team, customers) can certainly be seen as a special strength of Scrum. Scrum does not adapt and scale easily, which in recent years has led to the development of approaches such as SAFe which is based on Scrum at its core. Other PM concepts, such as *Disciplined Agile Delivery* (PMI) or *Agile PM* (Agile Business Consortium),[24] are likely to provide approaches to closing the gaps within agile methods (but outside Scrum) here – and will be the subject of further investigation.

---

21 in the present version 1.
22 see Boehm/Turner 2009, p. 35.
23 see SAFe 2019.
24 see Ambler/Lines 2020, respectively Agile Business Consortium 2019.

2.2.4   The process analysis in detail

Figure 2.8 documents the delta analysis at the level of the individual PM processes. The UPMF is used to answer the key question of whether or how well the respective PM process is covered by Scrum elements.

Legend:

- in Scrum not addressed (0)
- in Scrum implicitly contained (1)
- in Scrum explicitly included (2)
- in Scrum addressed in a particularly pronounced way (3)

Note: Even if this is not quite methodologically correct, numerical intermediate values have been entered in the ordinal scale where appropriate (if the assessment lies between two values) in order to ensure calculability.

2.3   Dealing with complexity

Managers want to believe that they can predict the future. Managers need to learn that predictability is impossible when complex, creative work [...] takes place.

(Ken Schwaber)

2.3.1   What does complexity mean?

The concept and the tracing of problems back to their complexity has been in vogue for several years. For example, a Google Internet search currently yields more than 9 million hits for the search term *complexity* and more than 27 million hits for the term *complex*. An absolute peak (4 times as many as today) was reached in September 2008, the month of the bankruptcy of the investment bank Lehman Brothers and the beginning of the escalation of the financial crisis.[25] Last but not least, there is the frequently used classification of the present time as the *VUCA-world* (see Introduction chapter), where the "C" stands precisely for *complexity.*[26]

The inflationary use of the term prompts us to take a systematic look at it. But what does complexity actually mean?

There is no uniform definition in the literature. Obviously, not least the colloquial use leads to a multitude of interpretations and thus variations. Everyone has a more or less intuitive understanding of the term. What everyone seems to have in common is that complex problems are always difficult to solve. This can be observed especially in the domain of PM. The still unsatisfactory success rates of projects with a simultaneous increase in the importance of project management are causing scientists and practitioners to search for new ways.[27] In many

---

25 GOOGLE Trends 2020 and SPIEGEL Economy 2020.
26 see e.g. t2informatik n.d.b.
27 see e.g. Standish Group 2018; GPM/ESB 2015.

ID	Process	Scrum?	Comment
[1.1.1]	Develop Business Case	0	PO ensures business value.
[1.1.2]	Carry out initial approvals	0	
[1.1.3]	Develop project order	0	
[1.2.1]	Identify Stakeholders	0	
[1.3.1]	Identify strategic risks	0	
[1.4.1]	Define Scope	3	A core element of Scrum is the use of a continuously evaluated and prioritised backlog.
[1.6.1]	Define strategic project process	1	Scrum specifies sprints as a process structure.
[1.7.1]	Plan project team	2	Scrum calls for an omnipotent, available, small, pulled-together team.
[2.1.1]	Develop a valuability strategy	0	
[2.10.1]	Implement configuration management	0	
[2.10.2]	Implement knowledge and document management	1	Knowledge is shared in Scrum primarily through communication; document management is not addressed.
[2.10.3]	Describe PM system (project)	0	Merely a demand not to dilute Scrum
[2.2.1]	Implement Stakeholders management	1	SHM is the implicit task of the PO, at least with regard to the product.
[2.2.2]	Set up contracting	0	
[2.3.1]	Implement risk management	0.5	Scrum is subject to the assumption that the incremental-iterative approach at least minimises product risks.
[2.3.2]	Analyse risks	1	No statement on environment-related risks
[2.3.3.]	Plan risk measures	1	Risk-oriented prioritisation of the backlog possible
[2.4.1]	Implement scope management	3	Active management of the backlog
[2.5.1]	Structure project work	3	Sprint Planning
[2.5.2]	Design organisational structure	1	Scrum role model
[2.6.1]	Create project schedule	2	Sprints
[2.6.2]	Shape PM system	0	
[2.7.1]	Design communication	2.5	In Scrum: Internal team and team-customer-communication; other stakeholders not explicit
[2.7.2]	Form project team	3	
[2.8.1]	Implement quality management	3	Reviews, *Quality by Design* claim *and Definition of Done*
[2.9.1]	Implement cost and financial management	0	
[2.9.2]	Implement resource management	1	Demand: Team available full time
[2.9.3]	Ensure resource provision	1	Demand: Team available full time
[3.10.1]	Administrate project system configuration	0	

FIGURE 2.8 Mapping UPMF-Scrum, details.

ID	Process	Scrum?	Comment
[3.4.1]	Raise requirements	2	User Stories
[3.4.2]	Ensure solution architecture & integration management	0	
[3.4.3]	Ensure transfer of operations	0	
[3.6.1]	Go through external testing procedures	1	If required as a backlog-item to be listed
[3.7.1]	Carry out project communication	2.5	
[3.7.2]	Ensure change management	1	Reviews make a contribution
[3.8.1]	Ensure test management	2	Sprint Review
[4.1.1]	Update Business Case	1	Continuous backlog-reprioritisation by PO
[4.1.2]	Evaluate project status	3	Short cycles, daily stand-ups, reviews
[4.1.3]	Control change requests	3	Continuous backlog adjustment
[4.1.4]	Issue phase releases	2	Reviews
[4.10.1]	Control knowledge and documents	1.5	Knowledge is shared through meetings and common artefacts; no statement on documents
[4.10.2]	Ensure project progress documentation	0.5	Scrum artefacts (charts) at least document the progress.
[4.2.1]	Involve the client	1	To be assigned to the PO
[4.2.2]	Manage Stakeholders	0.5	SHM is the implicit task of the PO, at least with regard to the product.
[4.2.3]	Manage contracts	0	
[4.3.1]	Monitor risk development	1	Reviews and implicit task of the PO
[4.3.2]	Implement risk measures	0	
[4.4.1]	Validate Scope	3	Continuous backlog evaluation
[4.5.1]	Update project structure	0	
[4.6.1]	Control work package processing	2	
[4.6.2]	Control project deadlines	2	Sprint clocking and continuous prioritisation
[4.6.3]	Update project plan	2	Sprint Planning
[4.7.1]	Control communication	0	
[4.7.2]	Steering the project team	2	Self-organisation of the team
[4.7.3]	Develop project team	1	Methodologically, cooperation through retros; professionally through the composition as an interdisciplinary team
[4.8.1]	Perform quality assurance	2	
[4.9.1]	Monitor costs and payment flows	0	
[4.9.2]	Controlling the use of resources	0	
[5.1.1]	Prepare project closure	0	Scrum tends to be continuous product development
[5.10.1]	Create final report	0	
[5.2.1]	Terminate contracts	0	
[5.2.2]	Determine project completion	1	Review last sprint
[5.2.3]	Transfer project results into operation	0	

FIGURE 2.8 (Continued)

ID	Process	Scrum?	Comment
[5.5.1]	Accept delivery objects	1	No statement on formal aspects; continuously in reviews, ensured by means of Definition of Done and Acceptance Criteria
[5.6.1]	Complete project phases	2.5	Procedures defined with content focus
[5.6.2]	Report project status	1.5	(only) Progress-related (burndown)
[5.7.1]	Finalise project organisation	0	
[5.9.1]	Return resources	0	

FIGURE 2.8   (Continued)

FIGURE 2.9   Dismantling and assembling a car: Complex? Or just complicated?

cases, agile procedures are identified as a suitable solution at this point.[28] But can this really be sustained?

2.3.2   Definition of terms

If one approaches the term systematically, it becomes clear that one must distinguish between the terms *complexity* and *complicacy*. Here is an example (Figure 2.9).

Assembling a car involves many interconnected elements. For a layman, this is a complex task. For an expert, it is just a complicated task that can be solved well by structuring, but redesigning such a system is always a complex task, because design decisions have an impact on other decisions. An example: Increasing engine power forces the strengthening of the braking system. So feedback occurs.

As a quintessence, although not a common denominator of the known publications on the subject, the following logical structure in the derivation of complexity can be synthesised:

28 see Komus et al. 2020.

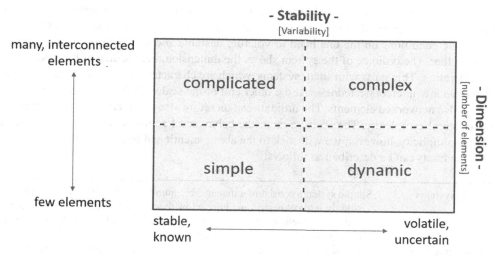

FIGURE 2.10   Cross of complexity.

A SYSTEM IS COMPLEX WHEN THE FOLLOWING FACTORS ARE PRESENT:

- There are many elements.
- The elements are networked.
- The elements have interactions.
- There is a dynamic that can be structural or in terms of content.
- This results in emergence.

This means that a complex system must first of all include many elements that do not exist separately from each other but are connected in some way. A large project with many work packages or a company-wide project portfolio with many projects can serve as an example. If such a system exists, then it can be described as complicated.

If the elements contained also show interactions that lead to feedback processes and thus to a dynamic in the system, then the complicated system becomes a complex one. This is usually the case with large projects and large project landscapes.[29]

In summary, such a structuring of the systems can be represented in a "cross of complexity" (see Figure 2.10).[30]

The quadrants are, of course, as usual in such representations, actually none, because the boundaries between the areas are fluid, basically even individual or organisation-specific. The example of car assembly may serve as an illustration here.

---

29 cf. Pommeranz 2011.
30 Own presentation based on Pommeranz 2011, p. 53.

The representation matrix is formed from the axes *stability* and *dimension*. The stability scales the willingness of the system to change over time: From stable, known and therefore quite safe conditions on the one hand to volatile, unstable and therefore uncertain conditions on the other. The ordinate of the system shows the dimension, i.e. the size of the system under consideration. This starts with small systems, which are characterised by few elements and thus also have low interconnectedness. At the other end of the scale are the extensive systems with many, also networked elements. The ordinate can therefore also be described as a representation of structural complexity. The abscissa maps the behavioural complexity accordingly.[31] For the sake of simplicity, however, we will stick to the above-mentioned terms from the diagram. The quadrants can be described as follows:[32]

Simple systems	Simple systems consist of a manageable number of elements that are only slightly interconnected and have a low dynamic of change. Due to the simple cause-effect relationships, both the structure and the future behaviour can be described and predicted in principle.
Complicated systems	Complicated (structurally complex) systems consist of a large number of elements that are interrelated, which makes them quite difficult to manage. Since they have a low dynamic of change, their future behaviour can be predicted.
Dynamic systems	Although simple dynamic systems can be described in terms of their structure due to their small number of elements and relations, they contain a significant potential for change due to their high dynamics. For this reason, a reliable prediction of future behaviour is hardly possible in detail.
Complex systems	Complex dynamic systems are defined both by a multitude of elements with multiple interactions and by an unclear potential for change. Due to this characteristic, they can neither be completely described in terms of their structure, nor can detailed statements be made about their future behaviour.

2.3.3 Procedure

The systems (e.g. projects) are as complex as they are due to their technical-content focus (Scope) or socio-economic context (organisation, environment). It is therefore a matter of finding adequate means of handling them. The premise is that a simple constellation is preferable due to the lower implementation risk and should therefore be aimed for. In Figure 2.11 this is expressed by the arrows.

The procedure for minimising the systematic or system-immanent risk can be described as follows in a symbiosis of diverse sources and experiences:

1. Structuring
    - Work Breakdown (→ Work breakdown structure, WBS)
    - Product Breakdown (→ Product breakdown structure, PBS)
    - Organisation Breakdown (→ sub-projects, organisational units)

---

31 see Pommeranz 2011, p. 53.
32 cf. Reinhard 2015, p. 147 ff.

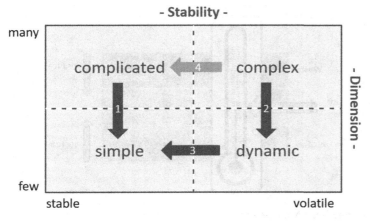

FIGURE 2.11 Procedure for minimising systematic risk.

2. Reduction
   • Prioritisation, focusing, e.g. through …
   • … scope reduction (→ smaller projects)
   • Release Planning (→ shorter project phases)
3. Iteration
   • Empirical process control
   • Rolling planning (→ Allow for imprecision)
   • Timeboxing
4. Stabilisation
   • Hypothesising to (provisionally) fix uncertainties (e.g. boundary conditions).
   • Big Design through Integration and Architecture Management

As already mentioned, simplicity is desirable and should at least be kept in mind as a guideline. Which path can and should actually be followed is ultimately due to the circumstances and the possibilities associated with them and must be shaped individually. Einstein provides his well-known wisdom on this:

Everything should be made as simple as possible, but not simpler.

(Albert Einstein)

2.3.4 Handling

In the sense of Einstein's quotation, the question of the right measure of (acceptable) complexity immediately arises. In principle, complexity is intrinsically difficult to measure. One measure recognised in cybernetics is *variety*. This is defined as the number of distinguishable states of a system.[33] The determination of this quantity proves to be of a rather theoretical nature,

---

33 Beer 1994, p. 247.

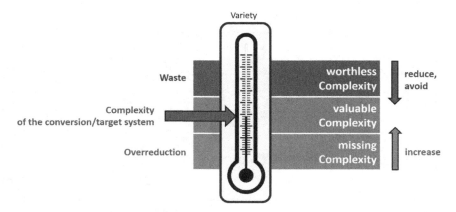

FIGURE 2.12 Ashby's Law and Lean Management.

as a simple example makes clear. Given n elements in the system, each of which can take on two different states. The number of states is therefore calculated as 2^n. With ten elements, this results in 1,024 states. With three possible states per element, this results in a variety of 59,049! How many work packages does your project have and how many states can they take on? In the project context, the number of communication channels with n team members can also serve as an illustration, which can in principle be calculated with n x (n − 1) / 2, i.e. it enters into the number of participants quadratically. Variety appears to be de facto unmeasurable in real complex systems in the field of management, but it remains a recognised thought construct that, with the help of *Ashby's Law,* at least offers help in designing systems: *The greater the (action) variety available to a control system, the greater the variety of disturbances it can handle.*[34]

In the context of Lean Thinking, this results in a scheme as shown in Figure 2.12 where the focus has to be on the project context.

The complexity of the PM system must therefore be large enough to handle the complexity of the project (at least equal in the narrow interpretation of Ashby's Law).

To simplify the constellation, adequate measures should therefore be taken as shown in Figure 2.11. If this is not possible then, due to Ashby's Law, mechanisms for controlling complex systems must be applied. In Figure 2.13 essential approaches are assigned to the corresponding quadrant – without claiming to be exhaustive.

Formation of networked subsystems:
The overall system is broken down into relatively self-sufficient subsystems. The subsystems work together in the sense of the whole and strive to maintain the underlying order. This construct can be recursive.[35]

---

34 cf. Heylighen/Joslyn 2001.
35 cf. Reinhard 2015, pp. 155–157.

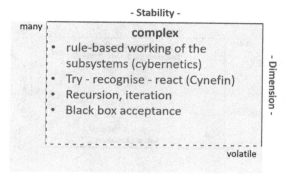

FIGURE 2.13   Mechanisms for managing unavoidable complexity.

Rule-based working:
The subsystems (inter)act on the basis of a set of rules. This makes it possible to react to unplanned constellations, at least in the sense of a heuristic, without knowing this situation *a priori*. Examples are traffic rules or a mission statement.[36]

Try – recognise – react:
Snowden describes this strategy in his *Cynefin framework*. Experimentation – for example in the form of prototyping – allows insights to be gained and practical approaches to be found. Here, the conditions for learning must be created, from which new practical insights then develop.[37]

Recursion, iteration:
This mechanic is implemented in particular through rolling planning. Here, specifying details too early is avoided. Instead, a planning status or concept is first roughly worked out and then taken up again at a later point in time ,close to the time of use, and processed further, renewed and detailed on the basis of the knowledge gained. This avoids waste due to planning errors and unnecessary planning effort.[38]

Black box acceptance:
Here, uncertainty is dealt with by accepting a certain lack of transparency. One orients oneself to the result instead of the processes, and the concrete procedure is not specified. In projects, for example, the work packages can be designed as such black boxes.[39]

2.3.5   The dilemma of complexity

Complexity has essentially been broken down into complicacy plus dynamics in this paper. In summary, one can observe – not least with regard to projects – a certain dilemma regarding the handling of complexity:

36 cf. Reinhard 2015, p. 177.
37 see e.g. Bayer 2010.
38 cf. Gessler/Sebe-Opfermann 2013, p. 73 ff.
39 cf. Gessler/Sebe-Opfermann 2013, p. 72, 75.

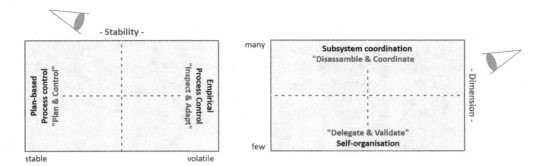

FIGURE 2.14 Basic guidelines for action.

Agile approaches are not well suited for complicated systems!
Boehm and Turner cite a few examples of project constellations studied and come to the conclusion that a plan-driven approach is better suited for large projects, indeed it is a must: *"Traditional plan-driven methods scale better to large projects. The plans, documentations and processes provide for better communication and coordination across large groups. [...] We see [...] absolutely no way to handle the problem with agile standup meetings and tacit knowledge propagation."*[40]

Plan-driven approaches are not well suited for dynamic, changing systems!
A Big Design Up Front – be it planning or technical design – in a volatile, changing system is doomed to generate waste. At a minimum, this includes delays and unnecessary work that has to be repeated at the time of implementation – if not errors caused by incorrect planning.

So it is in the combination of complexity and dynamism that the dilemma arises. As shown in Figure 2.12 it is therefore almost inevitable to try to move as far as possible towards the lighthouse of "simplicity". In particular, the observed effect of making projects large and extensive – and thus supposedly important – is counterproductive in this respect. *"Scaling up has proven difficult"* was already recognised by Boehm and Turner[41] and downsizing should rather be part of the instruments of skilful project design.

2.3.6 The quintessence of managing complexity

The considerations on the handling of complexity outlined above are generalised once again in the following and presented in their quintessence. To this end, the two spanning axes of the taxonomy used – stability and dimension of the system – are first considered individually and the guidelines for action are pointed out in the extremes with a view to application in projects (see Figure 2.14).

If the project situation is characterised by a (relative) stability, i.e. goals and ways of elaboration are known and relatively few changes are expected in this respect, then the procedure can and

---

40 Boehm/Turner 2009, p. 28, 29.
41 Boehm/Turner 2009, p. 28.

should be planned out by experts and the implementation of the plan should be controlled. This promises – with sufficient stability and competence – an efficient procedure that will not least lead to the goal. This approach can be summarised as *Plan & Control*, which corresponds to the so-called *traditional* project approach. If, on the other hand, the situation is assessed as volatile, this approach makes no sense due to the existing uncertainty or is at least a waste of resources, since a revision becomes necessary with every change. Instead, proceed in smaller steps, review the respective results and findings and then plan the next steps. This approach is generally not as efficient as a plan-driven one but allows for effectiveness appropriate to the unstable circumstances. This is also known as *Inspect & Adapt*.

If one turns to the size dimension of the project, then one should strive to make the overall system manageable through adequate division into subsystems, ergo sub-projects, work packages, etc. to make it manageable. Since it will hardly be possible to define sub-projects that are independent of each other (in this case it would be better to create independent projects), these will always have dependencies of various kinds on each other. The price for decomposition is therefore an increased need for coordination, which is expressed not least in systematic integration management and architecture design. The corresponding guideline should therefore be *Disassemble & Coordinate*. On the other side of the spectrum are the small projects. Small refers to the object of the project and/or the size of the team. The guideline here is the most far-reaching self-organisation possible, which avoids bureaucracy and promotes creativity and motivation of the actors in general.[42] Due to the low level of complexity, simple planning and working based on rules are often appropriate, which not least enables technical decisions to be made at the level of direct processing. Delegation usually also means that the results achieved are presented to the demand side for acceptance. Therefore, this guideline is summarised as *Delegate & Validate*.

The classification of a project into the constructed quadrants provides its complexity characteristic, which is always described by a combination of two values. Because of the complexity dilemma described, these guidelines must be combined in a meaningful way. From this the model shown in Figure 2.15 can be derived.

In the case of large projects that exhibit structural complexity due to a large number of interrelated elements, manageability should be achieved by breaking them down into smaller structures (XBS). This includes the creation of a work breakdown structure, a product breakdown structure and/or an organisational breakdown structure. The substructures created in this way must be reintegrated through systematic *solution architecture management*. In contrast to the structures described as (only) complicated projects in Figure 2.15, this reintegration must be emphasised even more in the case of complex projects, where a temporal variability is added to the structural size. In addition, the experience of very large organisations with extremely large project plans (e.g. the IT projects of the Federal Employment Agency, some of which are in the hundreds of millions),[43] shows that a release based approach is appropriate, in which a defined product bundle is developed and put into production on a quarterly basis (cf. also Programme Increment according to SAFe). For this reason, the procedure is referred to here as *release-based engineering*.

On the other hand, smaller projects also differ in their stability. Small stable projects are easy to plan; a simple, checklist-like schedule is often sufficient. Because of the simplicity, extensive

---

42 see e.g. Oestereich 2013.
43 see e.g. Beyer 2011.

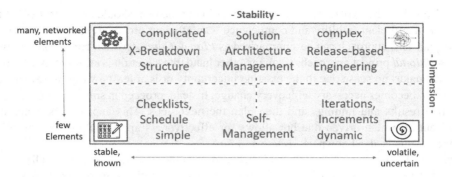

FIGURE 2.15 Generic guidelines for designing the project approach.

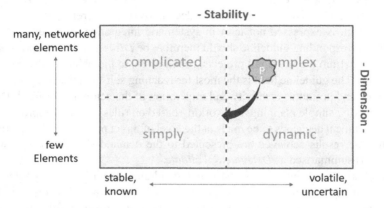

FIGURE 2.16 Classification of a project under consideration.

planning is not needed and would be a waste. The small but volatile projects must instead proceed recursively and learn from the development of one iteration step for the following step. The (product-oriented) checklist is managed here as a backlog that may need to be continuously adapted, if applicable. Scrum is the well-known process model for these small, volatile projects. The two poles of relatively small projects have in common that higher-level coordination is not absolutely necessary – neither in the form of solution architecture management nor as a distinct PM approach.

The explanations in this chapter provide a simply structured framework with which projects can be characterised in terms of their complexity from which an adequate project procedure and project management model can be derived. In doing so, one should always avoid projects that are unnecessarily extensive and large, and strive for security through the creation of stable conditions (see Figure 2.16). This ultimately leads to a hybrid approach in which, for example, through initial, agile project phases, frame conditions can be fixed and predictability achieved.[44]

44 see e.g. Blust 2019.

3

CORE PRINCIPLES OF LEAN PROJECT MANAGEMENT

After reading this chapter, you will know …

- the definition of Lean Project Management,
- the manifestations of waste within projects and a systematic classification in this respect,
- the adaption of the lean core principles to the domain of project management, and
- the description of the "3Gs" action maxims for Lean Project Management.

3.1 Definition of Lean PM

With the explanations in Chapters 1 and 2 the foundations for the design of the Lean PM approach are available. The task is now to integrate the two management domains – projects and lean – into an innovative new approach. We define Lean PM as follows:

DEFINITION LEAN PM

Lean PM is the extensive adaptation and application of Lean Management principles, methods and tools to PM processes and technical project delivery to increase the efficiency and effectiveness of project procedures.

In principle, it is often assumed that there is a trade-off between achieving efficiency versus effectiveness, i.e. both cannot be optimised (maximised) at the same time. Nevertheless, empirical studies show a positive correlation between these categories of performance.[1] For example, the

---

1 see Belvedere et al. 2019, p. 413.

DOI: 10.4324/9781003435402-4

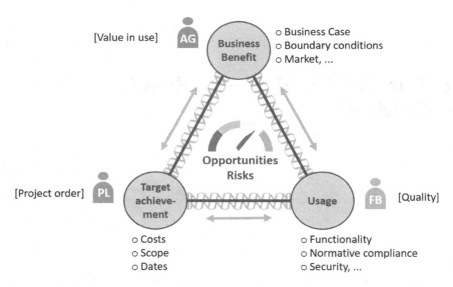

FIGURE 3.1   Value-based reinterpretation of the Magic Project Triangle.

early involvement of users in a project leads to the avoidance of waste (efficiency), but also to an increase in acceptance (effectiveness). Lean PM thus contributes to both categories, even if the term *lean* initially suggests a focus on efficiency.

With Lean PM, the focus of project work is strictly on the benefits that the project output generates (i.e. the outcome) – both individually and, above all, as a whole. This results in a holistic view of project management, in which the perspectives of the project manager are supplemented by those of the client (business benefit) and the user (professional use) in an expanded, value-oriented reinterpretation of the Magic Triangle (Figure 3.1). The project mandate is linked to the utility and the quality of the result.[2]

The cornerstone of the project assignment reflects the quite classical understanding of the goals and framework conditions under which the project manager is to complete the project. His or her immediate objective is still to complete the project according to the scope/performance, costs/budget and time/schedule (the classic Magic Triangle remains here). However, his/her view should be broadened in the direction of the project customers, users and clients. On the one hand, it is important to guarantee the use in the specialist area in accordance with the technical, functional and non-functional requirements. In a purely plan-driven approach, the requirements and functional specification which are considered formal order specifications, try to define this *a priori* and formally. The broader view should, however, allow for gains in knowledge in this regard in the course of the project, which is operationalised systematically and to the maximum extent in agile procedures. On the other side are the clients of the project, who can be internal or external. They have commissioned the project because they expect a business (economic, political or societal) benefit and attribute a positive utility value to the cost-benefit ratio of the

2 cf. also Kloss 2019, pp. 213–215 or Felchlin 2020, pp. 69–70.

project investment. This expectation is manifested in the business case of the project and may be subject to change over time due to changing framework or market conditions.

Inside the value-oriented Magic Triangle is the influence of opportunities and risks, which exert positive or negative pressure on the overall system. Depending on their occurrence, positive (when opportunities are realised) or negative effects (when risks occur) can result for all corners of the triangle.

EXAMPLES

1. A retail chain has commissioned the construction of another branch because the market analysis showed that there is an undersupply of the population in the district under consideration. In the course of construction, a similar branch is unexpectedly built in the same neighbourhood due to competition – the utility value changes. The project order will probably have to be adjusted and the sales concept of the branch may change.
2. In the course of the introduction of a standard software for human resource management (see Section 5.1), which lasted several years, the pay scale changed massively through collective bargaining. The employment groups were regrouped. This reduced the usability of the originally commissioned solution. As a result, the project order had to be changed and the value in use was reduced by the costs of the change.
3. The project work was delayed due to the effects of an unexpected flu epidemic, the go-live dates could no longer be met due to the absence of team members – the possibility of use was postponed and thus, if applicable, also the profitability analysis.

As in the classical Magic Triangle, all dimensions, i.e. the corners, are coupled with each other (see examples in the box). If one dimension changes, this usually has an impact on the others. The project manager, as the manager of the project, is usually paid for keeping to his project mandate (literally or only ideally), but in the sense of customer satisfaction he should also focus on the other dimensions of the Magic Triangle, and base decisions in the course of the project on them. For example, *Target Value Design* can be used, which is described in Section 4.3.3.

Like strict customer orientation, the application of the other lean values and especially lean practices lead to reinterpretations, be it in the sense of adapting of the lean elements or the classic PM elements. The characterisation of the lean elements according to values (paradigm and core principles) and practices (principles of action and methods, tools) provides a practical framework for this. The values represent the central principles of Lean Management, the implementation of which is what makes a concept a Lean Management concept. Questioning or abandoning them due to contextual considerations leads in principle to abandoning Lean Thinking as an overarching guideline. While this can be considered legitimate, it should then be called and classified differently as a management system. In this respect, Lean PM demands the literal application of the values and there is no need for any individual, project-related adaptation (see Figure 3.2). Lean designed projects follow the basic values of Lean Thinking![3]

---

3 cf. Erne 2019, pp. 70–71.

		Fundamental Application of Lean Elements	Contextual Adaption of Lean Elements
Values	Paradigms	X	
	Design Principles	X	
Practices	Acting Principles	X	X
	Methods, Tools		X

FIGURE 3.2   Framework for the design of Lean PM.

With the practices, on the other hand, elements of a more operational application level are present. It is therefore in the nature of things that these must generally be interpreted in terms of application. A prominent example of this is the Kanban system, which in its original context as a production Kanban has a quite different design than in the project context (see Section 4.5.2) – although the basic idea remains the same. It is certainly not wrong to say that the more operational the instrument, the more specific the immediate application benefit. In this respect, the figure shown in Figure 3.2 results in the need for contextual adaptation to project management. As will be shown later, however, a whole series of principles of action are so universal that they merely require implementation in the project context, not adaptation or even reinterpretation. Gemba serves as an example of a universal principle of action, Jidoka as an example of a principle that needs to be adapted (see Chapter 4).

APPLICATION OF THE LEAN ELEMENTS IN LEAN PM

The values of Lean Management (paradigm, core principles) are universal, i.e. they are seen as invariant to the application domain. The practices (principles of action, methods/tools), on the other hand, are to be designed in a domain-specific way. In doing so, opportunities for useful transfers arise in order to obtain innovative practices for PM.

The basic concepts of Lean Thinking must therefore be developed with reference to PM: *Customer* and *customer value, value stream, flow, pull* and *perfection*.

It can be assumed that a differentiating consideration between the PM processes in the narrower sense of planning and controlling the project on the one hand and the technical-progressive project processing on the other hand is relevant. In this context, the term *product* is therefore also characterised with reference to the project business. Likewise, in the first step, the concept of *waste*, which is fundamental for Lean Thinking.

3.2   Waste in projects

One way of identifying and classifying waste in projects is the one-to-one interpretation of the known types of waste in production, which have been so catchily formulated under the acronym

Tim Wood+ (see Section 1.2.2).[4] However, at this point we want to move away from the original *Tim Wood* classification somewhat in order to do more justice to the different character of production (processes) and projects.

Belvedere et al. carried out an empirical analysis in a company in the aerospace sector.[5] The processing of complex projects there, such as the development of a radio telescope, long-term data analysis for space missions or the technological development of coolers for satellites, usually takes several years and sometimes uses technologies that are only in the development stage. The results are complex "one of a kind" products of a one-off nature, but the development process is fundamentally repeatable and should therefore be carried out according to best practices. The analysis has identified a number of types of waste that cannot be characterised as specific to the cited study. Furthermore, the German Economic Production Working Group has worked out which types of waste are to be found in office work and has quantified them – non-representatively – as 38% of working time![6] The consolidation and generalisation of the above-mentioned analyses, supplemented by our own experience of various projects of different sizes,[7] yields the typical waste in projects described in Figure 3.3.

These elements of waste primarily focus on the administrative and controlling processes of the PM. They can basically occur in the forms of "too much" (e.g. reports that nobody reads), "too little" (e.g. missing documentation), "wrong" (e.g. unnecessary personnel changes) or "twice" (e.g. planning tools used in parallel). Depending on the type of project (e.g. plant construction projects), the known types of waste in production (Tim Wood+) also come into play in the technical project processing.

In the course of structuring the identified wastes to derive a typification, it becomes clear that they can be clustered by *type and area of waste*. In summary, the following project-specific areas of potential waste can be identified (see Figure 3.4).[8]

On the other hand, recurring patterns can be identified that typify the wastes according to their nature. These are already shown in Figure 3.3, noted with the assignment of the last column and are: Misallocation (A), unnecessary Movement (B), Defects (D), Overprocessing (O), Underprocessing (U), Waiting (W) and Misdirection (Z). Some of the well-known Tim Wood waste types can also be seen here. The specific character of projects, which is made clear by their uniqueness, complexity, teamwork, etc., is also reflected here and brings other relevant types of waste to the fore:

In the case of ***misallocation,*** personnel and non-personnel capacities, know-how, material, money or equipment are not adequately allocated to the project. A typical pattern is also harmful multitasking, which is known to lead to inefficiencies due to training or set-up times.[9]

---

4 cf. e.g. Brenner 2018.
5 see Belvedere et al. 2019.
6 AWF 2007.
7 These include various (major) projects, e.g. for the introduction of SAP or the introduction of a nationwide control centre system (see Chapter 5), but also a large number of project reviews carried out in international consulting (Scheer GmbH/Software AG).
8 A total of 161 wastes (according to literature research and from own experience) were identified and classified.
9 see Komus et al. 2016.

Waste	Type
Processes & Organisation	
Tasks with similar content are carried out in parallel without coordination. At project portfolio management level: Entire projects	O
Qualified new staff are not recruited according to need. E.g. processes for staff recruitment are too slow.	A
The staff budget is too small. Resources for staff recruitment are too scarce.	A
Processes are too bureaucratic. Either in the project or the complementary business processes, e.g. funds are transferred/released too slowly.	O
Complementary processes are too slow. E.g. procurement processes	W
Uncertain budgeting ties up project team and managers in acquisition activities. E.g. lack of funding leads to researchers and technicians being permanently in proposal mode.	A
Hardware is moved back and forth between different workplaces.	N
Projects are not completed.	A
The splitting of project activity across different locations leads to many transfers.	N
Unnecessary business trips take place. Physical meetings are preferred to virtual meetings.	N
Other process participants (people or machines) must be waited for.	W
Documents are transported through the buildings. (in-house mail, printer etc.)	N
Documentation & Data Processing	
Insufficient documentation leads to duplication of work. E.g. analyses have to be repeated because contradictory results are not comprehensible.	U
Duplication of work results from a lack of knowledge transfer. Lessons learned are not shared or not used.	U
There is an unnecessary amount of documentation.	O
Data is processed unnecessarily frequently. (storage, acquisition, transfer)	O
Data and information have to be searched for laboriously. Searching for documents, files, information of all kinds	N
Communication	
Information is not transmitted or not transmitted completely. E-mails do not contain all the necessary information. Follow-up of missing information, or unreachable colleagues, is needed.	D
Relevant information has to be filtered out of a flood of information. Emails are sent to people who are not responsible for the information reported. Sorting out over-information, junk mail, e-mail, spam, etc.	O
Decisions are postponed. E.g. decisions are postponed until milestones are reached.	W

FIGURE 3.3 Typical project waste.

Waste	Type
Meeting structures are inefficient.	O
E.g. too many meetings compared to real project needs, too many participants in project meetings, persistence in inefficient, too long or inconclusive meetings.	
Communication is insufficient.	U
E.g. also too few meetings compared to the real project requirements	
Competition between project teams prevents knowledge sharing.	U
Communication is not adequately structured.	D
Competence	
Staff deployment is inappropriate (over/under qualified).	A
E.g. the organisation of support services is delegated to professionals (travel bookings or the like) or the assignment of staff to projects is not done on the basis of competence but on the basis of availability.	
The project team lacks decision-making authority.	A
Lack of transparency about existing competences hinders the need-based staffing of projects.	A
Lack of a corresponding database etc.	
Staff turnover	B
E.g. young staff cannot be retained.	
Roles are poorly defined.	M
As a result, poorly delegated, unclear and confusing tasks have to be clarified, etc.	
Project participants lack commitment due to a lack of incentive system.	U
Commitment is not encouraged by the compensation system, etc.	
The staff is changed unnecessarily.	A
Service provision	
Tests are carried out without necessity.	O
E.g. due to misunderstood standards, unnecessary repetition	
Requirements are not adequately formalised.	D
Multitasking leads to unnecessary set-up and training times.	A
Work is unnecessarily extended.	W
(Parkinson's Law)	
Decisions come too late or are not implemented.	W
Incorrect or incomplete results must be corrected.	D
Sources of interference interrupt the flow of work.	W
Planning & Design	
Overspecification	O
Excessive requirements compared to real necessities, "golden taps"	
The specifications are changed frequently.	M
Sub-specification	Z
E.g. too general requirements	
The schedule is faulty.	D
Deadlines are unstable.	M
Milestones are often postponed etc.	
Planning that is too detailed requires constant revision.	O

FIGURE 3.3 (Continued)

FIGURE 3.4 Areas of waste in project management.

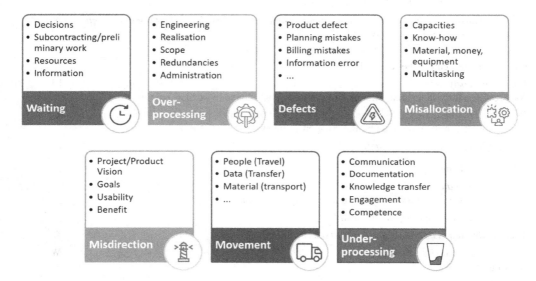

FIGURE 3.5 Types of typical waste within projects.

Also specific to projects is the risk of *misdirection*, where the project team mistakenly fails to work in the right or same direction. This includes poor project or product vision, unclear goals and a lack of focus on the usability and benefits of the solutions.

Finally, *under-processing*, which also does not occur in the production-related Tim-Wood waste types should be mentioned. Insufficient communication, documentation, knowledge transfer, commitment or competence leads to unnecessary friction, especially in projects that rely on these success factors.

In summary, the types of project-typical waste are as shown in Figure 3.5.

This identifies and typifies significant waste in projects, especially in the area of PM and project administration. When searching for waste in one's own projects, a helpful reference structure is available that can be used as a checklist.

The following chapter interprets the other key concepts of Lean Thinking in the project context. In the course of adapting the Lean Management approach to projects, the central terms will be defined.

3.3   Interpretation of the Lean Management core principles

3.3.1   *Customer and value concept*

Lean Management focuses on the customer and value creation in the sense of customer orientation.[10] In order to transfer the concept of the customer, the respective defining characteristics are first identified.

3.3.1.1   *Customer – General definition and characteristics*

The term customer is familiar and widespread in everyday business and personal use. Nevertheless, it can be observed that the term is also often applied to all kinds of contexts, a kind of fad – and this certainly true in an extended sense of the term. For example, in Germany applicants for unemployment benefit are referred to as clients by the job centres,[11] as are students at a public university.[12] But what are the defining characteristics of a customer from the perspective of service provision and process management? The following definitional elements can be identified:[13]

1. The customer is the **recipient of** a (desired) service (product, service, information).
2. The customer is (directly or indirectly) the **trigger of** the service delivery process (end-to-end view).
3. The customer **pays for** the service.

For the usual customer in a business context, it can be assumed that all three of these characteristics apply. However, how is it to be dealt with conceptually if one or two of these characteristics do not apply? For example, the welfare claimant does not pay for the service provision of processing his claim (3.). Also, the denial of funds, i.e. a negative decision, is certainly not to be seen as a (desired) benefit (1.).

In the further course of the conception, we assume that the customer characteristic is fulfilled within projects if at least one of the above-mentioned customer characteristics is fulfilled.

3.3.1.2   *Transfer of the customer concept to projects*

When attempting to transfer Lean Management approaches to project management,[14] authors often speak of a simply and directly applied customer concept.[15] That is, the customer is the one to whom the project result is handed over (usually at the end of the project). In my view, this approach falls short.

---

10 cf. Gorecki/Pautsch 2013, p. 20.
11 see BMFSFJ 2013, p. 1.
12 cf. Schwaiger 2003. p. 33.
13 cf. also Kirchgeorg n.d.
14 Project management as a generalising term for project business to avoid talking only about commercial projects, i.e. externally commissioned projects.
15 cf. e.g. Pautsch/Steininger 2014 or Grote/Goyk 2018.

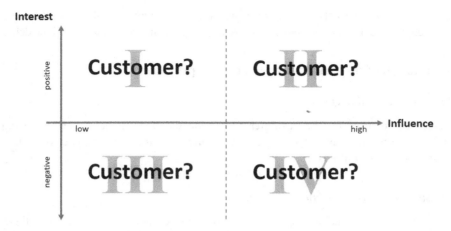

FIGURE 3.6   Stakeholder analysis for customer identification in projects.

In project management, the stakeholders of the project are referred to as groups of interest. "A stakeholder is a person or group that has a legitimate interest in the course or outcome of a process or project" (i.e. *interested parties*).[16] It therefore makes sense to use the stakeholder analysis familiar to PM as a method and starting point for identifying project clients. In the stakeholder analysis, the previously identified parties from the project and its environment in the company and beyond are classified in particular with regard to their influence on and attitude towards the project. This can be done in the form of a portfolio-like "force field" with these very dimensions (see Figure 3.6).

One possible interpretation would be to treat all identified stakeholders as customers of the project. In a way, this would be the antithesis of the narrower interpretation of many authors as described above. However, this approach does not lead to the desired results and is not very practicable: Stakeholders who have little influence on the project and who also have a negative attitude towards the project (for example, competitors in a product development project) are certainly not customers – not even in an extended view (1, 2 or 3 do not apply). The question now arises whether customers can be identified in a simple and unambiguous way from this force field representation.

Stakeholders who meet characteristics 1, 2 or 3 are necessarily in quadrant II, they have a high level of influence (they pay) and they have a high level of interest (they receive benefits from the project). On the other hand, these stakeholders from quadrant II do not necessarily fulfil the customer characteristics – e.g. a political interest group in a major construction project – but it is nevertheless critical to success and in this sense value-creating to treat these stakeholders in a customer-oriented way.

But project customers can also be identified in other quadrants: For example, it often makes sense to also involve those affected by the project results who may have a negative attitude to the change associated with the project (quadrants III and IV) – e.g. later users – in

16 Eilmann et al. 2011, p. 71.

a customer-oriented way. Thus, no direct rule can be derived from the stakeholder analysis of a known kind that identifies the project customer. We therefore derive the following defining characteristics for the concept of customer in the project context:

Customers in the sense of Lean PM are stakeholders who …

- obtain a service from the project (directly or indirectly) (cf. 1.),
- have commissioned a service provision (internal or external) (cf. 2.) or
- (formal or informal) have a high influence on the course of the project or the acceptance of the project outcome (cf. 3.).

Customer characteristics can also be determined depending on the (project) process under consideration. In order to shed more light on this aspect, the question of value streams in the project arises.

3.3.1.3 Identification and characteristics of the value streams in the project

The Lean Management concept essentially includes the alignment of value streams with the creation of added value for the customer. As previously stated, the customer of a project is not only the customer in the narrower sense. In this respect, the value streams of a project are also diverse and are identified and characterised in the following. The term *value* itself can be used in the usual understanding of process and Lean Management: Value is a service that is of use to the customer and for which the customer is generally willing to pay a price.[17]

The starting point is the general understanding that views a project in terms of PM processes and technical project processing (PP, project procedure). "Creating Value" means fulfilling the project mandate. The project mandate, in turn, should follow a business case, which describes the overriding value of the project in the form of business benefits.[18] Here, a reference to those responsible for the project can be presented: The direct measure of value for the project management and the project team is the project assignment; the measure for the project client is the business benefit. In this respect, there is an immediate classification in the project level model (see Figure 3.15).

What is the aim of the PP level?

Z1	The aim of project processing at the PP level is to produce a technical project result that meets the benefit/purpose-related requirements.

What is the goal of the PM level?

Z2	The aim of the PM is to ensure that Z1 is successfully achieved within the frame conditions set by the project brief.

---

17 cf. Gorecki/Pautsch 2013, p. 23.
18 cf. Office of Government Commerce 2017, p. 46 ff.

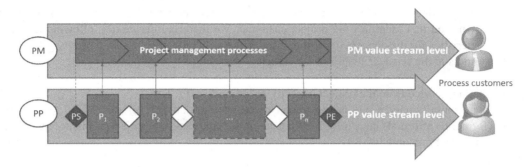

FIGURE 3.7 Value streams in the project.

Thus, the degree of concretisation and detail of the project order induces the PM system of the project. Contexts that are characterised by uncertainty thus lead, for example, to agile, iterative procedures with empirical-adaptive project control, while others are plan-based in specifications and project plans (see Chapter 6). New approaches to projects, especially agile methods, also break away from the classic client-project manager understanding, for example by introducing a role such as the product owner. However, we assume a broader understanding of PM that includes, for example, elements such as risk or stakeholder management and retains the general, fundamental and systemic view of projects with a separation in principle between project provider and project client.

The project portfolio management level represents the client's point of view. The aim of this level is to ensure that the benefits of the project as part of the project landscape are ultimately conducive to the higher-level goals of the organisation in the sense of a cost-benefit analysis (see Chapter 8). At this point we will first focus on the individual project level. The value generated by the project processes is therefore that which contributes to the achievement of the identified goals Z1 and Z2. Thus, two (three) value stream levels of the project are to be distinguished: Project management (PM) and project procedure (PP) (and project portfolio management (PPM)) (see Figure 3.7).

As previously explained, the project customers are to be identified in the course of an extended stakeholder analysis and ultimately be evaluated in terms of their importance and handling. Taking into account the two identified fundamental value streams in the project, project process-specific customers generally emerge:

The customers of the PP processes may be different from those of the PM processes.
Customers must be identified on a process-specific basis.

Typical clients at the project approach level are:

- Client
- User/applicant
- Follow-up process(es) within the project (incl. PM processes)
- Product and/or Process Owner

Typical clients at the PM level are:

- Client
- Team
- Stakeholder depending on context
- Product or process owner
- Company process(es) outside the project that have an interface (sink) to the project

3.3.2   Flow- and pull-principle

Make value flow is one of the core principles of Lean Management.[19] This principle calls for value creation processes that are not interrupted by the storage of intermediate or end products and by waiting times in the production process.[20]

What are flows in projects?

The flows in projects are to be located on the two levels of the value streams – PM and project approach. In this respect, the analysis also requires a two-part approach.

3.3.2.1   Flow in the technical-progressive project processing

At the project approach (PP) level, the primary added value of the project is developed. The operational units of the development are classically the so-called *work packages*, (PMBoK Guide, PM4, PRINCE2), which require the detailed processing of specific tasks. (In this respect, the structures of agile procedures such as Scrum characterised by work units such as Sprint, user stories and tasks). The flow of development is represented in the project flow charts of the phases or sprints of the project. In this respect, it is important in Lean PM to design these processes without significant and, in particular, unnecessary interruptions.

However, projects generally do not allow for a *One Piece Flow* according to which the flow can be aligned but are by definition characterised by a certain complexity: A multitude of activities, dependencies, delivery objects and participants that need to be coordinated. With Stalk & Hout's Golden Rule (see Section 3.3.4), Parkinson's law (extension of work)[21] and Goldratt's bottleneck theory,[22] a rule can be derived, at least heuristically, which is already used in the well-known *critical chain project management approach (CCPM)*.[23] The (optimisation of) processing procedures in the project should be based on the following criteria:

- after the bottleneck resource (e.g. the only remaining developer who still knows the ancient programming language);
- the (causally) critical path while simultaneously
- avoiding the inclusion of safety buffers in the work packages;

---

19 Bicheno 1998, p. 7.
20 cf. Gorecki/Pautsch 2013, p. 22.
21 see Parkinson 1955.
22 see Goldratt 1990.
23 see Techt/Lörz 2015.

- the parallelisation of activities makes sense here, as long as harmful multitasking of resources[24] is avoided as far as possible.
- Furthermore, activities of the project approach (i.e. primarily value-adding) are to be prioritised over those of the PM in case of conflict or bottleneck.

3.3.2.2 Flow in the PM processes

The PM processes represent the area of the project in which the recurring activities of secondary value creation take place (planning, reporting, change processing, etc.). Flow here means providing results promptly when they are needed (the plan must be up to date), making decisions quickly (otherwise the project waits) or actively managing project synchronisation points such as milestones and quality gates or reviews (Scrum) so that no avoidable and unproductive waiting times occur. The milestones in particular already show the conflict, because the alternative should not be to work without synchronisation elements, because otherwise duplication of work may occur in the further course of the project.

The derivation for the design of the flow in PM can therefore be identified as follows:

- Establish short decision-making channels, if applicable using a (temporary) bypass;
- Encourage communication between staff so that clarification can be achieved quickly if there is a time lag;
- Clear, clean, timely timing in project reporting, consisting of status reports and *jours fixes*;
- Timely feedback and action derivation as required by the project status (report);
- Timely provision of resources
  and
- Project planning according to the principles of flow in the project approach level (see above).

3.3.2.3 How can the pull principle be implemented in the project?

The pull principle states that the value stream is primarily set in motion by the customer's need or demand. Therefore, production only takes place when the services are needed.[25] However, this must be understood in such a way that the entire demand chain is considered, not just the direct environment or even the immediate downstream process. In many cases, the latter is even impossible: "You cannot grow an orange tree overnight to provide a pulled orange drink."[26] Forward planning is therefore not to be rejected in general, but the consequence is that the planning and production cycle should adapt to the demand cycle. If sales are daily, production should be daily.

Current developments in the IT sector, for example, postulate that a smooth and continuous transition between development and production should be guaranteed (*DevOps*).[27] The motto here is "Develop in Cadence – Deliver on Demand", i.e. develop in a fixed rhythm and

---

24 see Techt/Lörz 2015, p. 41.
25 cf. Gorecki/Pautsch 2013, p. 22.
26 Bicheno 1998, p. 7.
27 see Alt et al. 2017, p. 23. The term *DevOps* refers to the cooperation between software development and IT operations.

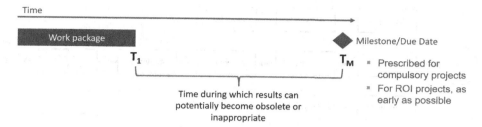

FIGURE 3.8 Timing of work packages.

productively implement the results according to the requirements of the business. This rhythm must be aligned with the requirements, i.e. as a rule it must contain the shortest and most timely cycles possible.

But let's move on to the transfer to project management.

3.3.2.4 What does the pull principle mean for the project approach level?

At the level of professional-progressive project work, pull means ...

Backward scheduling:
In backward scheduling, the logical project schedule is planned starting from its desired end date. Analogous to production and material planning, the timely determination of the (technical) result requirements takes place – while guaranteeing the latest possible deadlines so that no unnecessary gaps occur.[28] In the sense of the pull principle, this reduces the time in which results can potentially become obsolete or unsuitable (see Figure 3.8).

Critical Chain Project Management has already taken up this idea. Here it also becomes clear how to deal with (time) safety buffers, which are still indispensable in good project planning: Systematically applied, centrally managed, bundled at the end and reduced to a minimum (see Figure 3.9). The risk of planning errors in terms of time – both positive and negative – is inherent in the one-off nature of projects.

Work in Progress Limitation:
The CCPM also postulates, in particular, the avoidance of harmful multitasking.[29] One possible practice for implementing this principle has developed from the Kanban method of Lean Production: *Kanban boards are* used to organise the work flow in projects in a self-controlling way. Such a board, which can be specifically designed depending on the project context, is in principle as shown in Figure 3.10.[30]

The pull principle is implemented to the extent that when the processing capacity is free (i.e. the current workload is below the so-called *work-in-progress limit (WiP limit)*), the next pending task is pulled by the processor/team. Here, therefore, a capacity-related pull mechanism is implemented, in contrast to the demand-related pull in the Lean Production system.

28 see Wiendahl 2014, p. 322 ff.
29 cf. Techt/Lörz 2015.
30 see e.g. Timinger 2017, p. 202.

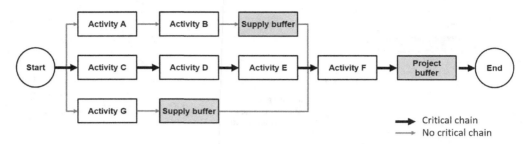

FIGURE 3.9   Critical chain and buffers in the CCPM.

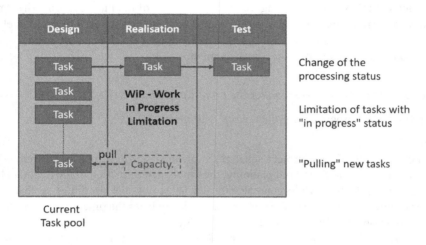

FIGURE 3.10   Project Kanban Board.

3.3.2.5   What does the pull principle mean for the PM level?

At the level of the PM processes, *Pull* would mean that information (because this is what this domain is primarily about) is generated and made available when it is requested. For example, the project manager asks his team for a status report, which is then produced. However, best practices such as PRINCE2 go in a different direction and call for *management by exceptions*, i.e. reporting according to the push principle (only) when there is an appropriate reason.[31]

In my view, however, both are impractical in practice, as delays as well as missing and incomplete reporting are inevitable. Therefore, in practice, a reporting system based on firmly established and – ideally – cyclical reporting adapted to the project-specific requirements, such as the duration of the project, has also become established. The decisive factor in the sense of the pull principle is that the rhythm is appropriate to the project and that the recipient of the report also processes the information received in a timely manner. Specifically: A project that lasts three months should have

---

31 cf. Office of Government Commerce 2017, p. 24 ff.

at least a weekly status report, a (large) project that lasts three years can generally allow for longer intervals. The overarching *heartbeat* of the organisation – such as monthly steering committee meetings – also pulls at the information delivery in this regard.

Professional PM is characterised by *proactive action* – i.e. a risk analysis is better not done when the crisis has already occurred; communication with stakeholders not only when they are already creating difficulties, etc. In this respect, the pull principle is not generally applicable here.

But one important message remains:

> Do not produce (PM) results in a stockpile, but in a timely manner, taking into account the latest findings, oriented to the follow-up action!

The situation is different with the management of general information and existing and newly generated knowledge. Here, in the sense of Lean PM, it makes sense to implement access according to the fetch principle in order to avoid a so-called *information overflow*, in which information (partly untargeted, flood of mail) is scattered without need (time, content) from the recipient's point of view.[32] There is a need for adequate (project) knowledge management, or at least document management:

> General project-related and overarching information (knowledge) should be provided according to the fetch principle!

3.3.3 Perfection

Striving for perfection – this is a central core principle of Lean Thinking.[33] But what does perfection mean in the project context? Can it be the overall goal of a project to produce (only) perfect solutions?

The project mandate provides the basis for the question of how much perfection is to be achieved in a project. This is the link between the overarching project benefit (business case) and the concrete, if possible *smartly* defined operative project goals.[34] The project order often includes (as an annex) a requirements specification and a functional specification. In agile approaches, the requirements and solution approaches are usually only concretised in the course of the project (and the project goals are defined less "*smartly*").

The well-known *Magic Triangle,* consisting of performance, costs and deadlines, supplemented by the aspect of client satisfaction,[35] provides the context: You cannot push performance and also satisfaction in general to the maximum if the time and budget for this are not given. Perfection in the project context can therefore only mean: Maximum efficiency

---

32 cf. Pautsch/Steininger 2014, p. 113, 148.
33 cf. Womack/Jones 2013, p. 111 ff.
34 *smart* is a well-known acronym for specific, measurable, acceptable/applicable, realistic and scheduled.
35 cf. e.g. Ottmann et al. 2008, p. 89.

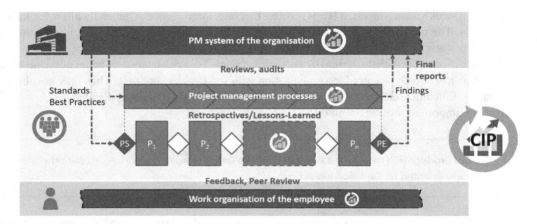

FIGURE 3.11 Levels of the project CIP.

in the project processes to achieve the required customer benefit within the framework set for the project. There is also a connection to quality management, because here too it is not about maximum ("gold edge solutions"), but about required quality.[36]

The origin of the Lean principle of striving for perfection was already formulated in 1996 by Womack and Jones in the context of Lean Production (coming in particular from the automotive industry). Here, the focus is primarily on recurring processes, which are therefore sensibly subjected to a CIP (Kaizen).[37] Projects are by definition (however) unique in their respective characteristics, so they are not actually subject to repetition. Nevertheless, processes can of course be identified that are neither unique nor carried out once. These include in particular the (entire) PM processes. But also the activities of the project procedure level, namely the technical elaboration of the project result, often show characteristics of repetition: Depending on the project type (e.g. IT development or construction project), standard and best practice procedures provide input and at the same time the addressee for continuous further development. Examples of such standards are the German V-Modell XT (IT projects)[38] or the PM services in the construction and real estate industry according to AHO,[39] which describe in particular – but not only – the technical project procedure.

This ultimately leads us to consider the CIP on two levels. Firstly, within the project itself and secondly, at the higher level of organisational PM, i.e. the PM system of a company. In addition, the self-organisation of the employees should not be disregarded (see Figure 3.11).

Depending on the level of consideration, various typical methods of CIP arise. At the project level, Lessons Learned should be systematically identified, as has been established, for example, in the Scrum in the form of retrospectives. The decisive factor for the immediate project benefit

---

36 cf. Becker n.d., p. 2.
37 cf. e.g. Womack/Jones 2013, p. 33.
38 see Höhn et al. 2008.
39 Committee of the Associations and Chambers of Engineers and Architects for the Fee Regulations e.V., cf. Preuß 2014.

is that these retrospectives are not only carried out at the end of the project, but regularly during the course of the project, for example at phase transitions (in Scrum: *Sprint transitions*). Classical PM standards address this requirement, such as PRINCE2 with the principle of *learning from experience* in the process of *managing phase transitions*.[40]

From a longer-term, higher-level perspective, there is a demand for continuous development of the PM system, i.e. the pursuit of perfection in organisational PM. Project findings that can be generalised should be incorporated into the PM system via project completion reports – and generally documented in a PM manual.[41] The knowledge about projects that is documented and institutionalised in central organisational units such as the PM Offices or project portfolio management can then be transferred into the PM system and flow *a priori* into the projects in the form of company-specific standards and at runtime in the form of supervision (reviews, audits).

Finally, there is the human being as an individual team member, whose individual work organisation – especially under the aspect of project teams that tend to organise themselves – contributes decisively to the success of the whole. In addition to self-reflection, feedback from the supervisor and within the *peer group*, i.e. in the project team "among equals", is a typical instrument.

In addition to *kaizen*, Lean Management also knows *kaikadu*, or radical improvement.[42] This is an alternative way of improvement through radical redesign of processes, such as is also known from business reengineering.[43] Possible reasons for a rigorous approach can be, for example:[44]

- Due to changes in the personnel structure (e.g. the dissolution of a PM Office), certain actors can no longer participate. "Business as usual" is therefore no longer possible.
- The qualification of the acting actors is not sufficient to fulfil the intended tasks.
- People who are supposed to implement certain requirements/tasks are not (no longer) available and transferring those requirements to other roles seems too difficult.
- The maturity level of the established processes is no longer adequate, for example due to changed environmental framework conditions (e.g. the requirements of digitalisation in the industrial environment), so that fundamental changes in the procedure must be implemented.
- Customers of a process have changed or the company processes have established changed regulations (e.g. the new head of controlling demands a different project reporting system), so that the project processes have to be fundamentally changed.

An obvious example for kaikadu is the introduction of project portfolio management in a company where projects have neither been carried out according to defined standards nor has the project landscape been systematically managed in its entirety. Accordingly, the task of organisational change management is vital in order to establish the radical change beneficially in the organisation, i.e. to anchor it sustainably.

---

40 cf. Office of Government Commerce 2017, p. 246.
41 cf. Seidl 2011, p. 152 ff.
42 cf. Womack/Jones 2013, p. 33, 112 ff.
43 cf. EABPM 2014, p. 510.
44 following Fleischmann et al. 2011, p. 170.

To sum up: Striving for project perfection means striving for maximum PM and project procedure process efficiency (= avoidance of waste) and for optimal effectiveness through the achievement of the direct and indirect project goals (= value creation of the project). In the context of projects, perfection can be concretised by values such as speed, adherence to schedules, effort efficiency, customer and employee satisfaction as well as effectiveness of results, etc. In general, all levels of project management are affected – the organisation, the project itself and the employees. Systematic project knowledge management is the key to improvement.

3.3.4 Project products

In the course of adapting of Lean Management to projects, the concept of product must also be considered specifically. In Lean Production the concept of products is obvious, but in project management a differentiated understanding arises, since project products refer not only to the technical domain but also to the management domain (cf. also Lean Administration).[45]

Project products are generally also called outputs or deliverables.[46] It becomes clear that the term *project product* includes all those achievements that are produced in the project as intermediate or final results. In general, the application goals are achieved through the development of technical products, while PM-related products, usually documents or information, are also generated to achieve the execution goals. Figure 3.12 schematically shows a typical classification of project products using the example of IT projects.

The delivery objects in the different branches of this tree generally have different typical characteristics, such as versioning, release or quality assurance processes.

The technical products as well as the technical (final) documentation are deliverables that are always directly or indirectly required by the project order; intermediate results and PM documentation result from the necessity of project execution. In this respect, a Lean PM perspective results in a general prioritisation of the former (technical products and functional documentation) – at least in the case of conflict. However, all types of products contribute in principle to the achievement of the project's goals and thus to its added value. Superfluous products in this respect, which are required, for example, by a company's PM system that is too general or restrictive, should be eliminated (cf. also the management principle of *scalability and adaptability* according to PRINCE2).

Thus, in the sense of customer-focused added value orientation, a prioritisation of the project products from I to IV results consistently (see Figure 3.12). This does not mean that the delivery objects of category IV, i.e. the project system documentation, e.g. the quality assurance plan, are not value-adding and thus superfluous. Prioritisation merely provides a recommendation for action in the event of a conflict (such as a delay in deadlines). In general, the following should apply: Direct value-adding activities and products take precedence over (only) indirect value-adding ones, which can also be classified as process-related or business-related waste (see Section 1.2.2). Stalks & Houts *Golden Rule* provides practical help:[47] "Never delay a value

---

45 cf. Brenner 2018, p. 6 ff.
46 cf. Gessler 2011, p. 329.
47 Stalk/Hout 1990.

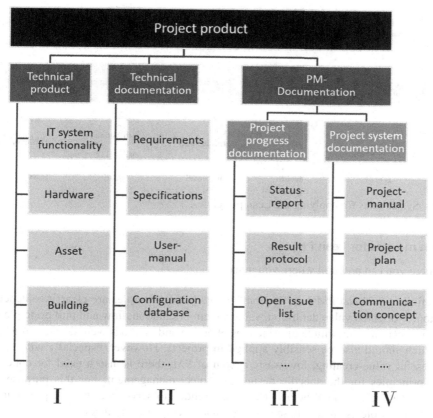

FIGURE 3.12 Project-product type tree.

adding step by a non-value adding (although temporarily necessary) step", which can also be applied in the context of project management.

3.3.5 Application of the Lean principles

With the identification and characterisation of project-typical products, another building block is now available for operationalising the lean PM approach as a customer-oriented and product-related PM framework. All in all, with regard to the beneficial application of Lean principles, the results are summarised in Figure 3.13.

In order to achieve the benefits of Lean PM – to increase effectiveness, avoid waste, increase project throughput, achieve load distribution of resources, in particular of employees, and demand equity of activities, as well as to increase the overall efficiency of project work – the Lean principles are applied. However, in order to be able to apply them, the prerequisites must first be created by identifying and naming the process customers, the service provision processes, the relevant flow objects and – last but not least – the performance targets for the projects under consideration. Then the Lean principles can be applied in concrete terms.

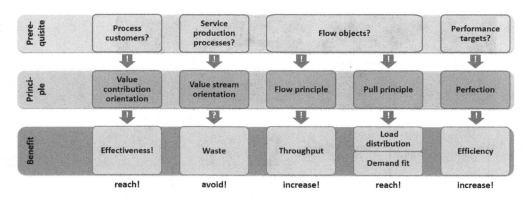

FIGURE 3.13 Systematics for applying the Lean principles.

3.4 Action maxims for Lean PM

Flow where you can and pull where you must![48]

With the definition of Lean PM (see Section 3.1), its basic principles are clear: Avoidance of waste, customer-oriented value definition, value stream orientation, flow and pull principle and striving for perfection. A multitude of principles of action and methods were derived from this, many of which should also be sensibly applied in projects. However, especially with regard to the successful value-creating, low-waste design of PM, there is also a need for concisely formulated principles for the specific design of project management. For this purpose, the *3Gs for Lean PM* are presented below. They are intended to serve as a guiding principle for successful, "lean" projects and are based in particular on empirical findings.

3.4.1 Selected well-known guiding principles for action

In the past, a number of requirements have already been published, which in the following have served as suggestions for the *3Gs of Lean PM*. These include, for example, the Agile Manifesto which has its origins in software development, but also the eight principles of Leach, the twelve principles of Pautsch and Steininger or the five principles of Erne, all of which were developed under the heading of Lean PM. All of these guiding principles are to be evaluated as practicable aids for the implementation of projects and are more or less directly oriented towards the basic principles of Lean Management. Let us therefore take a look at the maxims for action that have been presented by the various authors in the past.

Pautsch and Steininger, for example, have developed twelve principles, which are as follows:[49]

1. The management of the pre- and post-phase of a project must follow the same principles as for the actual performance of the project itself.

---

48 Ballard et al. 2007, p. 151.
49 see Pautsch/Steininger 2014, p. 114 ff.

2. The value of the project outcome is the benchmark for managing and building and aligning the drivers of the project.
3. The value stream is designed with a view to reducing or eliminating waste.
4. Suppliers and project partners with a significant contribution to the project result must be seamlessly integrated into the value stream.
5. The project management is not only administratively active but fulfils a real management task.
6. Determining the method of project implementation is the PM's responsibility.
7. Formulating a convincing vision of the project outcome is necessary for project success.
8. Establishing a lean project culture is a decisive factor for project success.
9. The communication of the project status and the relevant key performance indicators is carried out according to the principles of Visual Management.
10. Lean PM requires a balance in the use of project resources.
11. Project services are created when they are needed.
12. The pursuit of perfection drives the engine of project innovation.

Erne, on the other hand, develops five principles in the sense of transferring the general Lean principles to PM. These are:[50]

1. Specify the optimal benefit-effort ratio from the customer's point of view.
2. Define the minimum value-creating work packages and work processes.
3. Establish clear responsibilities, tasks and competences at the lowest possible organisational level.
4. Ensure a continuous flow of results by limiting work in progress.
5. Identify errors immediately and eliminate them sustainably.

Leach published eight Lean principles for project success as early as 2005, which can be summarised as follows:[51]

1. Project system: Define an effective system for the environment and the projects
2. Leading people: Ensuring that stakeholders actively support project success throughout project implementation
3. Commissioning: Establish a project vision and agree a project brief
4. Right solution: Provide the right solution to the problem or opportunity that the project wants to exploit
5. Managing variation: Understanding variation and introducing appropriate management for buffers and risks
6. Project risk management: Develop measures to reduce the likelihood and possible undesirable consequences of identifiable risks
7. Project plan: Provide all project stakeholders with a bottleneck-oriented resource-based schedule and appropriate execution procedures
8. Execution: Enable relay team performance by helping the team decide which task to work on. The focus is on identifying and rewarding the achievement of project success.

---

50 see Erne 2019, p. 91 ff.
51 see Leach 2005.

Previously, the Poppendiecks also provided seven Lean principles in the context of Lean Software Development in the implementation of Lean Thinking for (IT) projects:[52]

1. Avoid waste
2. Support learning
3. Decide as late as possible
4. Deliver as early as possible
5. Giving responsibility to the team
6. Build in integrity
7. See the whole

With the *3Gs guiding principle,* these approaches are now to be expressed in summary form and enriched not least by many years of own project experience as well as empirically proven project success factors.[53]

3.4.2 The 3Gs for Lean Project Management

In an effort to write a concise guiding principle for the practical implementation of Lean PM that accompanies the project practice, the "3 Guidelines" (3Gs) for Lean PM are formulated as follows: *Participation, Pareto-orientation* and *Fit* (see Figure 3.14).

These 3Gs stand for a series of principles for successful PM. They can be used to design a hybrid PM in the sense of Lean Thinking. Modern plan-driven and also consistently agile project procedures can be integrated. The three principles of this leitmotif are briefly described below.

Guideline 1 Participation
The first principle calls for the comprehensive implementation of the idea of involving the stakeholders of the project as intensively as possible, but above all continuously during the implementation of the project. The central stakeholders are the customers of the project. This includes in particular the later users of the project results as well as the clients (Figure 3.15).

The principle of Lean Management, which is to only develop services for which there are customers who benefit from them and are (potentially) willing to pay for them, can only be implemented in a targeted manner if these customers are involved. Experience also shows that stakeholders who are not direct customers but who have a regulatory influence (referred to here as regulatory institutions) – for example, by having to approve results – should be involved at an early stage in order to increase their acceptance. An example of this is the right of participation of a works council.

> Downstream stakeholders should be involved in upstream acivities.

---

52 see Poppendieck/Poppendieck 2003.
53 see Hüsselmann 2020, pp. 61–71.

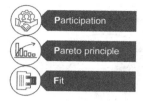

FIGURE 3.14   3Gs for Lean PM.

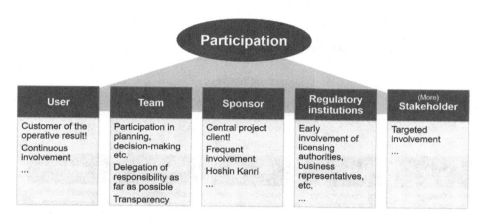

FIGURE 3.15   Principle of participation and its elements.

This sentence, based on Ballard, figuratively describes the demand for active participation of the respective process customers in the course of the project, as Figure 3.16 illustrates.[54]

In sequential process models, as is often the case in construction projects for technical reasons, one can often say: The devil takes the hindmost! Those who are involved with their trade or sub-project in the later course of the project have to "pay for" what was produced in the preceding phases of the project – often at least in the sense of catching up with delays. In order to achieve smooth transitions, fewer errors, better communication, improved know-how transfer, less time loss and familiarisation effort, etc., it is necessary to demand that the process customers of one's own work (i.e. the subsequent trades) are always involved in the production of one's own trade, i.e. in the flow of value creation the downstream project teams participate in the work further upstream.

Guideline 2 Orientation towards the pareto principle

The Pareto Principle – often referred to as the *80–20 rule* – states that the cost-benefit ratio is optimal within a certain performance range. More performance (output), such as maximised quality, can only be achieved with often disproportionate additional effort. A rule of thumb says that 80% of the result can be achieved with 20% of the effort (see Figure 3.17).

54 Following Ballard 2012.

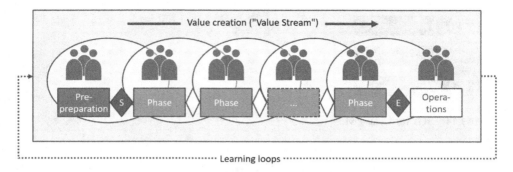

FIGURE 3.16 Participation of subsequent stakeholders.

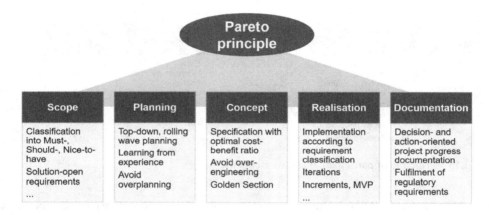

FIGURE 3.17 Pareto principle and its elements.

This insight should be used in projects in different areas, as shown in Figure 3.17, for example, by classifying the requirements according to the scheme "must have" vs. "should have" vs. "nice to have" in order to achieve an optimal cost-benefit ratio in the realisation and to remain flexible-adaptive. Over-engineering and over-planning should also be avoided, as the dynamics of a project can easily lead to waste in these areas. Other aspects are a rolling, top-down oriented planning of the project process, which in the strict design of a purely agile procedure completely dispenses with the details of the next but one phase (see Section 4.4).

NOTE

The pareto principle is (only) at first glance contradictory to the core Lean principle of striving for perfection. Here, perfection can only mean: A perfect cost-benefit ratio in the sense of focusing on purely value-creating activities and carrying them out with as little waste as possible. The pareto optimum is a perfect constellation in this respect.

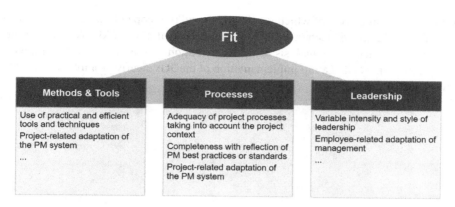

FIGURE 3.18   Principle of fit and its elements.

Guideline 3 Fit

The third overarching principle is what I call the Fit, by analogy with technology. The management and the management system must fit the specific framework conditions of the project (see Figure 3.18).

In many companies, PM is carried out according to standards and best practices. For example, this includes the ICB or the work PM4 of the German Association for Project Management. These and other bodies of knowledge claim to be comprehensive descriptive compendia for PM. In application, however, these PM frameworks absolutely require a company-specific and ultimately project-specific adaptation. Otherwise, project administration quickly becomes bureaucracy without any particular added value.[55] Worse still, aspects that are really important under the conditions of the project may be neglected or suffer from a lack of acceptance, e.g. risk management (see Chapter 6).[56]

But the management style of the project leader must also adapt to the personality traits of the team members – and not vice versa. Some employees need maximum freedom to develop their potential, others prefer clear tasks and instructions. In this respect, the postulate of a principally pure self-organisation of teams is rejected here and rather a situational, albeit tendentially delegative, approach to this issue is called for (see Section 7.1.3).

The 3Gs of Lean PM are intended as a guiding principle to help implement Lean PM in practice. They are not intended to replace the general principles of Lean Management but are to be understood as a specific design for projects and assistance for PM to achieve value creation while avoiding waste. Basically, the 3Gs summarise the generally documented findings from research and practice and are validated and tested through many years of own project experience. Of course, they do not claim to guarantee project success, but they can provide an impetus to better achieve this goal. In a meta-study to identify empirically proven success factors 14 studies from 2008 to 2017 were evaluated in the Lean PM working group of the Universities of Applied Science in Koblenz & Central Hesse (Germany). 122 success factors

---

55 cf. Erne 2019, p. 29.
56 see Hüsselmann et al. 2019.

were identified, the enumeration of which would go beyond the scope of this article. However, they can be discussed in connection with the publication of the UPMF (Unified Project Management Framework).[57] The analysis of the contribution of the 3Gs to the fulfilment of these success factors shows that a full implementation of the 3Gs clearly promotes at least 85% of the known success factors.

---

57 see Hüsselmann 2020, p. 61 ff.

4

PRACTICES OF LEAN PROJECT MANAGEMENT

After reading this chapter, you will know …

- the characteristics of practices within a management approach, and
- a lot of selected acting principles, methods and tools for PM disciplines like contract design, scope management, project planning, process control, etc.
- the "Agilometer" – a tool for project classification regarding agile or plan-driven approach, and
- the project value stream analysis for process-oriented design of project work.

4.1 What are practices?

A key function in the application of Lean PM, but also of management approaches in general in day-to-day operations, are domain-specific practices. According to the Capability Maturity Model Integration, practices describe activities that should be applied to achieve the goals of a process area.[1] A distinction is made here between specific and generic practices. Generic practices are basically applicable to several process areas, but usually have to be specified in the application for the respective process area. The Project Management Institute (PMI) defines a specific type of technical approach as one that contributes to the implementation of a process and uses one or more methods or tools as a practice.[2]

Practices, as understood in this book, serve to operationalise the fundamental principles (values) of the management approach. For example, a Kanban system enables the practical implementation of the Pull principle. In this respect, the practices focus on the "*how*" and "*who*" of a concept, while the values justify the "*why*" and "*what*". In the Lean PM context, we count generic principles

---

1 see Hertneck/Kneuper 2012, p. 15.
2 see PMI 2017, see 713.

DOI: 10.4324/9781003435402-5

of action among the practices, such as *Gemba* ("Go to the place of action!"), as well as concrete methods and tools, such as the Ishikawa diagram for root cause analysis. We therefore define:

PRINCIPLE OF ACTION

Generic practice that represents an operationally usable guide to action for the implementation of fundamental values – without specifically specifying the form of application. Examples: Gemba, Kaizen, Poka Yoke etc. They can be further operationalised through methods or tools, but do not have to be in order to be usable.

METHOD OR TOOL

Specific practice that is an operationally useful instrument in solving a task type within the concept. The task may occur in several contexts, e.g. root cause analysis and risk analysis, but the practice will usually be primarily useful to one task type. Examples: Failure Mode and Effects Analysis (FMEA), Makigami, Kanban etc. Tools are methods that are equipped with specific artefacts. Example: 8D report, Ishikawa diagram. A sharp demarcation is not always possible (but may also not be valuable).

The systematic analysis of the principles of action described by a large number of authors for Lean Management in general results in the following top 10 as a sort of common denominator (see Figure 4.1).

The data shown in Figure 4.1 have their origins in Lean Management, or more precisely in Lean Production. However, it is easy to see that, regardless of the systemic differences between process-oriented and project-oriented work, all ten principles can be transferred one-to-one to project management in a beneficial way.

In contrast to the clearly defined values of an approach, in the case of Lean Management the central paradigm and the core principles, the number of practices increases significantly in general. Depending on the contextual circumstances, practices are to be used specifically according to the contextual requirements. It is by no means necessary to demand that all available practices, especially methods and tools, are used. On the contrary, in Lean Management in particular, the set of available practices is basically impossible to define conclusively. What makes sense is what helps – and the striving for continuous improvement of the system sets no limits to the imagination here. Nevertheless, during the period of application of Lean Management, a number of practices have emerged that have proven to be particularly useful.

At this point, it is worth mentioning once again the connection between Lean PM as an overarching guideline for the design of a (hybrid) PM system and the Unified Project Management Framework (UPMF) as a process-oriented reference model for PM. The UPMF can be used as a company-internal standardised process model for the analysis and design of PM

Action principle	Description
Team spirit	Tasks are completed as a team. The focus is on achieving the goal together. Internal competition is avoided.
Personal responsibility	Each employee is responsible for the fulfilment of his or her own tasks but can fall back on the team at any time.
Feedback	Each activity is followed by feedback to identify successes and areas for improvement.
Standardisation	For a common consensus, work steps are standardised both in writing and visually.
Immediate defect handling	... at the root. Every employee is involved in the quality process. Process disruptions and errors are analysed immediately (in the team) until the cause is identified and eliminated.
Visual representation	... of information, errors and problems. Complex interrelationships, goals, targets, standards as well as material and process flow are visually prepared. Errors are visually presented in order to draw attention to them. (Andon)
Anticipatory problem solving	For zero-defect quality, possible errors are already avoided before they can occur. Anticipation instead of reaction. (Poka Yoke)
Small steps for improvement	Improvement takes place in small steps and not in radical changes. One improvement is made on the feedback of a previous one. (Kaizen)
Problems at the scene	... to be analysed. Problems and process disturbances are analysed at the place of occurrence and, if applicable, remedied in order to be able to recognise possible effects of a decision. (Gemba)
Culture of failure	Accept mistakes in the form of self-reflection, learn from them and develop strategies to avoid them in the future. (Hansei)

FIGURE 4.1 Top 10 action principles of Lean Management

processes. For this purpose, the processes of the UPMF are orchestrated to the relevant value chains as in a modular system, the areas of waste are identified with its help and the processes and organisational structures, optimised according to the principles of Lean PM, are designed (see also Section 1.1.2). This approach can thus be classified as a practice of Lean PM.

In the following sections, some further selected principles of action and methods that are used in the context of project execution to implement Lean Thinking are presented. The selection includes, in particular, practices that have been used for some years in construction projects in the context of Lean Construction, as well as others that have been successfully applied above all in the company's own IT and organisational projects. Attention was paid to the transferability to basically any type of project and the transfer was achieved by generalising possible type specifics.

4.2 Integrative contract design

4.2.1 Characteristics

The ideals of Lean PM are particularly suited to successfully managing commercial projects. The orientation towards customer benefits in the handling of the project, not least the end customers for the project result, is the most obvious evidence of this. The management of commercial projects, also called *Commercial Project Management*, requires not least a distinct contract management. From the drafting of the contract to the monitoring of the contractually owed performance, from claim management to the formal acceptance and handover of the project results, the commercial accounting and the subsequent handling of warranty and guarantee services, this is a separate domain of the PM with a thoroughly legal character.

In the bilateral design of project contracts, the following basic distinctions can be made at the commercial and performance level:

Performance: Contract for product delivery versus contract for services
Commercial: Fixed price contract versus contract on a time and materials basis

---

In a *contract for product delivery,* the contractor owes the client a successfully delivered product for which he receives remuneration (turnkey project). A typical example in the context of a project is the technical system to be built according to the agreed specifications or functioning software. The core of a *service contract* is – as the name suggests – the provision of a service as opposed to a service in kind. An example in the project business is the development work of an engineer. What is typically owed is the provision of work in accordance with the current state of the relevant technical domain (state of the art). The term *fixed price* usually means a lump-sum price which is intended to cover all individual services required to produce a contractually compliant and defect-free service. In the case of a contract based on *time and materials,* on the other hand, the client agrees to pay the contractor according to the time and effort actually incurred. A lump sum price is often not seriously calculable here due to the existing uncertainties regarding the expenses, at least on the part of the contractor.

With the basic classification described – a further differentiated consideration of the different respective variants etc. is left to the lawyers here – it becomes clear that contract design is not unimportant in the implementation of Lean PM. Even if contracts for product delivery are often paid for with a lump sum and services are paid for on a time & material basis, it is a widespread misconception that only these combinations are possible. By separating the legal aspects (what is owed?) from the commercial aspects (how is it billed?) it is also possible to conclude contracts for product delivery that are billed according to time and material. In many cases, of course, the billable expenditure is capped so as not to create a bottomless pit. In order to distribute risks and opportunities between the contracting parties, appropriate control mechanisms must be agreed upon, such as tolerances, threshold values, bonus-malus rules, etc. The decisive factor here is that (legal) regulations, like regulations on acceptance, transfer of risk, warranty and, if applicable, guarantee, must be preserved by the character of the contract.

4.2.2 The relational contract

Lean PM principles such as learning from experience, open error culture, participation etc., however, require other aspects of adequate contract design. It is about the way of cooperation in the execution of the project. Classic contracts for project implementation are so-called *transactional* contracts. They clarify the object of performance, including all rights and obligations of the project participants, through regulations that are as detailed as possible, justiciable and secured with liability sanctions. The contract is intended to anticipate possible (negative) developments in advance, distribute risks and secure claims and liability. This often leads to a competitive mindset among the project participants, with little incentive for joint value creation orientation, but rather a focus on individual economic interests (silo thinking) and little collaboration.[3] In the construction industry, which is characterised by commercial projects with a large number of partners to be contractually involved – depending on the constellation, client/principal, project planner (at various levels), project controller, project manager, construction company and subcontractor/subcontractor for various trades – the often problematic consequences of classic contract design have led, since the 1980s at the latest, to the worldwide development of contracts designed as partnerships, the so-called *relational* contracts. These have emerged as reactions to typical, structural problems in a large number of construction projects: Construction delays, unilateral risk shifting, discovery of unworkable plans only in the execution phase, hidden price competition including the awarding of services at prices that are not economically adequate with the consequence of a high number of supplements, etc. The transferability of these facts to other types of projects, such as large software implementation projects, plant construction, etc. is obvious.

Starting from the construction industry in Australia, some forms of relational contracts have been developed in the Anglo-Saxon world since the 1990s. These are usually multi-party contracts that are geared towards (only) joint project success and are characterised by the following elements:

Cooperation and collaboration as the basis for project implementation

- Integrative regulation of the service provision processes (conception, realisation etc.)
- Joint development and responsibility for project goals
- Early involvement of key stakeholders
- Avoidance of information breaks and asymmetries
- Values such as openness, transparency, trust, honesty, reliability, willingness to learn, mutual support and Lean Thinking

Definition of mechanisms to promote cooperation

- Common organisational structures
- Spatial cooperation
- Unanimity, at least as a principle
- Joint project management
- Integrated IT system for the technical and administrative control of documentation

Solidarity-based incentive system incl. risk and opportunity community (Shared Risks)

- Win-win or common loss strategies
- Joint Reward
- Joint risk bearing, at least partially (if applicable graduated according to participation)
- Transparency (Open Book)

---

3 see Bethan 2020.

Joint risk management
- Monetary reserve for unforeseen cases
- Use of risk register (tool)
- Early warning system
- Shared responsibility

In summary, relational contracts shape the relationship of the project partners in such a way that the responsibilities and benefits of the project are shared fairly and transparently. In doing so, the mechanisms of implementation are based on trust and partnership. This generally leads to improved working relationships between project participants, efficiency and effectiveness of management and technical project processes, conflict resolution capacity and ultimately economic success (for all).[4] In many cases, they are long-term contracts that develop over time and make lasting demands on the relationships between the partners. Over the past 20 years, several forms of relational contracts have emerged in the construction industry worldwide, e.g. alliance contracts, Project Partnering Contract (PPC), Integrated Form of Agreement (IFOA) as well as model contract conditions such as the New Engineering Contracts (NEC) or ConsensusDOCS.[5]

4.2.3   The agile fixed price contract

One variant of contract design is the *agile fixed-price contract*.[6] The aim of this form of contract is, among other things, to retain the character of a contract product delivery for the reasons mentioned above, but at the same time to achieve significantly increased flexibility in the provision of services. In the sense of a lump sum price, a monetary budget is agreed upon, which, in the case of project types dominated by personnel costs, such as software development, ultimately leads to a contingent of man-days for project implementation. The other side of the coin is the scope of the contract, i.e. the scope of the commissioned work in kind (system, IT, etc.). A rigid set of requirements or specifications provides little or no leeway here. In order to achieve possibilities for management, methods such as those described in the section on "breathing" scope and rolling planning should be used.

4.3   Breathing scope

Making the commissioned service more flexible also requires making the agreed scope of the project more flexible. Instead of a rigid scope that has already been finally defined by the project contract, a *breathing scope* that can be changed within limits should be used. The term is explained below, which also includes the *MuSCoW rule* and the *Target Value Design/ Engineering*.

---

4 see Colledge 2005, p. 1.
5 see Heidemann 2011, p. 39.
6 Opelt et al. 2014.

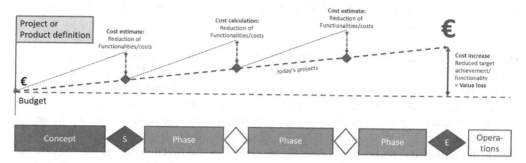

FIGURE 4.2   Frequent course of the development of a project.

4.3.1   Definition of the breathing scope

The definition of the scope of a project, i.e. the scope in depth and breadth, is a central task within the project initialisation and thus the project order. Here it is determined which functionalities (e.g. introduction of an ERP logistics module) in which organisational context (e.g. all producing national companies) in which design (e.g. automated supplier integration), if any, are the subject of the project. The simpler the project context and the more stable its conditions, the more binding the scope can be defined. The more complex and thus temporally unpredictable the project object is, the more a rigid scope will hinder the positive, benefit- and cost-oriented development of the project. Figure 4.2 shows a not infrequently encountered cost development of a project that must be avoided.

The task is therefore to design scope management – not least in customer projects – in such a way that the PM has meaningful scope to lead the project successfully to its goal. This motivates the idea of a breathing scope in the style of a current account, for the realisation of which two important methods are presented below. The *MuSCoW system* for the predominantly functional-technical prioritisation of requirements is supplemented by *Target Value Design* which is taken from Lean Construction projects and which extends the prioritisation to include the benefit-cost ratio. Combining these two methods results in innovative scope management in the sense of value-added-oriented Lean PM.

4.3.2   The MuSCoW system

In order to achieve flexibility in the processing of the project scope, it is important that the corresponding performance catalogue of the functionalities (functional specification) or requirements (specifications), which in large projects can contain several thousand items (see Chapter 5), is prioritised. In many cases, however, it can be observed that such service catalogues lack such prioritisation, which, for example, in customer projects that are carried out as fixed-price contracts for work and services regularly leads to extensive change requests that increase costs, delay completion, postpone completion dates and, last but not least, tie up management capacity in contract or claim management that could be used elsewhere and more effectively in the project. Such unprioritised service catalogues can often be observed in public tenders. The *MuSCoW system* provides a practicable rule for prioritising the items in the project scope. *MuSCoW* stands for

Must-have, Should-have, Could-have and *Won't-have*, which classifies the individual items of the requirements catalogue.

Must-have requirements are technically indispensable for the solution. Without their implementation, the solution is of no use – it simply does not work. They therefore define the minimum requirements for the solution, such as the product. The acronym MUST also stands for *Minimal Usable SubseT*.[7] The product-related subset of this is also known as *Minimum Viable Product* (MVP). Its implementation is basically non-negotiable (example: The brakes of a car).

Should-have requirements are not directly necessary for the functionality of the solution but have significant relevance to the achievement of project benefits. If they are missing, there may be short-term remedies. If resources (time, money, capacity) are available, they should be implemented to ensure customer satisfaction (example: The automatic transmission of a car).

Could-have requirements (also called nice-to-haves) can be excluded more easily, as they are only relevant to the general functionality of the solution to a limited extent. If they are missing, there may be simple and also permanent remedies. In the course of the project, they represent the most obvious bargaining chip in the management of the scope and serve primarily as swap items in the evaluation and scheduling of change requests. However, it should not be completely disregarded here that these nice-to-have elements can make an important contribution to customer satisfaction as *enthusiasm factors,* as the *Kano model* shows.[8] (In this respect, the must-have elements can be roughly compared with the so-called *basic factors* and the should-have elements with the *performance factors* in terms of customer satisfaction).

Won't-have requirements are not implemented in the actual project planning. However, it is optional to list them in the catalogue of services and at least offers the advantage that they can be referred to in the event of possible shifts in priorities and/or possibilities. Furthermore, they serve to actively delineate the expectations of the project, what will be implemented and what will not.

Figure 4.3 shows the logic of the MuSCoW system with a view to the active scope graphically. The Minimal Viable Product cannot generally be equated with the MUST, since the performance catalogue of a project must also include items that have nothing to do with the pure product – such as services. Of course, this applies in particular to projects that do not serve product development, e.g. organisational projects.

A scope is created in the logic of a current account, i.e. developments in the project, especially changes, can be credited or debited to the scope account. With the definition of corresponding tolerances (cf. overdraft framework), there is room for manoeuvre for the operative PM. In agile approaches, the *project backlog* (i.e. the requirements catalogue) is even prioritised and designed on an ongoing basis and change is planned as an immanent part of project management. This

---

7 see Agile Business Consortium 2019, p. 70.
8 see e.g. Jochem 2019, pp. 57–61.

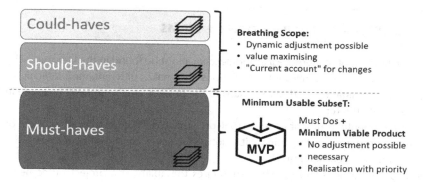

FIGURE 4.3 Breathing scope with the MuSCo(W) rule.

seems difficult, at least in external customer projects that are based on a contract, especially if it is a fixed-price contract.[9] Here the MuSCoW system offers a practicable framework – and is thus independent of the question of whether it is an (internal) investment project or an (external) customer project. The MuSCoW system proves to be practicable and helpful, especially in the case of customer projects (see Section 5.2).

4.3.3 Target Value Design

Target Value Design (TVD) was developed a few years ago for construction projects.[10] It is an adaptation of the target costing approach of the stationary industry to projects. Target costing places the desired market price and thus the customers at the centre of the development. It uses a retrograde calculation to answer the question of how much a product may cost. In this way it differs from the traditional cost calculation, the *cost-plus calculation* – i.e. *cost + margin = offer price*. Target costing is an attempt to realise customer orientation with regard to customer-specific product features or functions. It is therefore particularly suitable for the development of complex products that are manufactured in medium batch sizes.

Target Value Design takes up this idea. It extends the classic Magic Triangle to include the aspect of the value of solution elements for the customer. The following example illustrates the difference between purpose (sense/benefit) and project mission (or project objective):

Purpose:	Reduction of the time for the journey from point A to point B
Project mission	Construction of a bridge (cable car, ferry etc.)

In the following, the core elements and the application of Target Value Design are explained and applied to any type of project.

---

9 cf. Opelt et al. 2014.
10 see Ballard 2012.

VALUE AND BENEFIT – A CLARIFICATION OF TERMS

A uniform understanding of the terms *value* and *benefit* cannot be found in the literature. The often English-language origin of relevant publications also makes differentiation difficult, as the term *value is* predominantly used. With regard to the project, an attempt will be made in the following to clarify the terms:

An *impact* fundamentally describes an actual or expected change through a new or changed system compared to the initial situation. The *value contribution* describes the monetary valuation of a recorded impact.

The *benefit*, on the other hand, generally describes the ability of a project outcome to satisfy certain needs of one or more stakeholders, e.g. the job-to-be-done. The impact of the project is thus not only captured from a monetary point of view.

Value in use is the degree to which the (ideal) benefit for a stakeholder is achieved. We look at the extent to which the project result is subjectively, monetarily or non-monetarily suitable for the satisfaction of needs (benefit).

In a broad interpretation of the term *value creation* in the context of Lean PM, we thus refer to the value in use, in a narrower interpretation to the (monetary) value contribution.

Target Value Design begins in the initiation and preparation phase of a project, in which the framework conditions for the project are defined and lead to the project mandate. Based on the purpose of the project, the first step is to identify the value for the customer, i.e. the benefit. In the case of an infrastructure project, for example, it can be determined whether cost optimisation, prestige (lighthouse project), sustainability, functionality, return on investment or employee loyalty are more important to the client, i.e. whether they represent a primary value. For example, the construction of the Elbe Philharmonic Hall in Hamburg has shown that although the costs have increased tenfold (handling problem), the result is assessed as a success for the client due to its reputation (application success).

The values are then transferred to functions of the solution spectrum. In the value analysis, the functions are prioritised by the client (e.g. by pair comparison).

The value analysis is followed by the (rough) cost calculation for the elements, e.g. functionalities, various possible solutions. Here, in the sense of *set-based design* no early restriction of alternative solution possibilities should take place (see Section 4.4.2). The costs are determined and the associated solutions are prioritised again. The result is a cost-benefit matrix (see Figure 4.4).

The prioritisation that can be taken from this matrix represents the basis for real value engineering: Thus, there are binding cost specifications with a controlling character for the team already at the beginning of the project in the product development phase, which can be significantly influenced in each stage of the project. The determination of preferences weights the costs against the importance of the various product features and it can be determined in the course of the project whether a product component or function may be overdeveloped or whether there is still potential for value enhancement. The cardinal rule of Target Value Design is: The target costs of the solution can never be exceeded! This means in implementation:

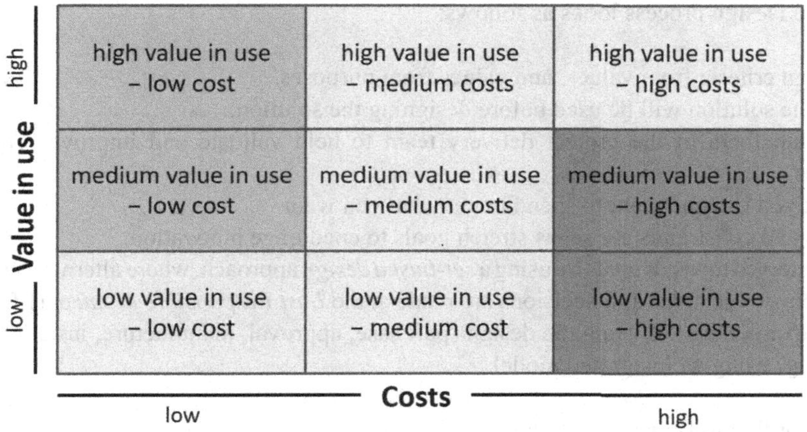

high	high value in use – low cost	high value in use – medium costs	high value in use – high costs
	medium value in use – low cost	medium value in use – medium costs	medium value in use – high costs
low	low value in use – low cost	low value in use – medium cost	low value in use – high costs

Value in use (vertical axis)

Costs

low   high

FIGURE 4.4   Cost-benefit matrix.

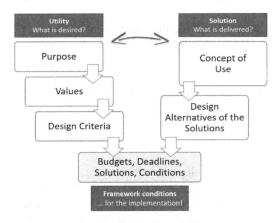

FIGURE 4.5   Target Value Design in the project definition (Project Definition Process).

- Ensure that whatever target costs increase somewhere in the solution, costs elsewhere are reduced by an equivalent amount without compromising application and quality.
- The refusal to grant additional scope to a project that exceeds the target cost.
- Managing the transition from planning to implementation to ensure that target costs are never exceeded.

Figure 4.5 shows schematically the procedure for the definition of the project with the help of Target Value Design.[11]

11 Following Ballard 2012, p. 15.

The Target Value Design process looks as follows:

- Develop design criteria from values, and values from purposes.
- Clarify how the solution will be used before designing the solution.
- Engage key members of the project delivery team to help validate and improve project business cases.
- Indicate what you can and want to spend to get what you want.
- Target values and constraints are set as stretch goals to encourage innovation.
- The design is steered towards goals by using a *set-based design* approach, where alternatives are evaluated against objectives and decisions are made at the *Last Responsible Moment* (LRM).
- Users create instructions for using the design (purchase, approval, manufacture, installation, commissioning) using an integrated model.

In summary, according to Ballard, it can be said:[12]
   Target Value Design ...

- ... is a management practice that steers the project to deliver value for the client and develops the solutions within the project constraints.
- ... requires a fundamental shift in thinking from expected costs to target costs.
- ... necessarily requires cross-functional teams. No single person has all the knowledge.
- ... requires an integrated product/process/cost model.

   With Target Value Design, the value-oriented Magic Triangle of PM (see Section 3.1) can be operationalised.

4.3.4 Weighted Shortest Job First

The *Weighted Shortest Job First* rule (WSJF) was developed by Reinertsen and is used, for example, within the SAFe concept.[13] It is a variant of the determination of a comparative cost-benefit ratio of a task. The WSJF is often used in the context of prioritising backlog items, not least at the level of the features of a solution that is being sought or further developed (cf. SAFe). It relates the opportunity costs to the effort of creation and is calculated as follows:

Weighted Shortest Job First = Opportunity Cost / Job Duration

With the opportunity costs, the WSJF indicator thus takes the approach of including the costs of a delay, i.e. of refraining from development. These are calculated as the *sum of the value in use for the customer plus time criticality plus risk reduction* (value of new findings):

---

12 see Ballard 2012, p. 3.
13 cf. Mathis 2016.

Value in use:	(In the original) Business value, e.g. increase in turnover or reduction in costs.
Time criticality:	Influence of a certain deadline on the value, e.g. customers dropping out, which may make the completion of the task worthless.
Risk value:	Opportunity for risk reduction and chances; riskier tasks should be tackled earlier than less risky ones, according to Reinertsen

These variables are each expressed on a Likert scale, with values roughly from 1–10, as is the time taken to complete the job (usually related to staff effort). In this way, for the various tasks available for selection in the backlog the WSJF value is determined, which leads to a ranking of the task completion. The following scheme then results for the process of application:

- Estimate the size of the backlog items.
- Estimate the values of the backlog items.
- Calculate the WSJF index values.
- Plan and control your workload according to the WSJF prioritisation.

Due to the construction of the WSJF indicator, it is easy to establish a connection with the Target Value Design. Both methods make use of the value in use that a feature represents for the customer. In this sense, the cost-benefit ratio that results from Target Value Design and the prioritisation of positions derived from it could be called *Valued Cheapest Job First* (VCJF) – the utility value for the customer is divided by the expected costs of a feature. As this change of name makes clear, the focus of the distinction is on the one hand on the costs (VCJF), and on the other hand on the duration of the realisation (WSJF). Which variant should ultimately be used in the project therefore depends on its subject. For example, the market launch of a new product certainly requires a strong focus on the time component, while an internal organisational project probably primarily has the costs in mind (even if this may represent an excessive reduction of the facts).

4.4   Goal-oriented, flexible project planning

4.4.1   Last Responsible Moment

Deciding too late is dangerous – but so is deciding too early in a fast-changing world!
(adapted from Jeff Atwood, American software developer)

The Lean Construction Institute (LCI, University of California) coined the term *Last Responsible Moment* (LRM) in the early 2000s. This was transferred to software development by the Poppendiecks, so that a further application domain, Lean Software Development, was added. The Lean Construction Institute defines the Last Responsible Moment as the moment when the cost of delaying a decision exceeds the benefit of the delay, or the moment when not making a decision eliminates an important alternative.[14] The LRM principle thus aims to avoid waste

---

14 see LCI 2017.

by committing to a solution alternative too early, by reducing the likelihood of revising already achieved results associated with a wrong decision if the commitment turns out to be wrong as knowledge is gained on an ongoing basis. Typical decisions in this context are architectural decisions, the implementation of which entails a lot of detailed work. If decisions are kept open as long as possible, this leads to *set-based design* from the perspective of the required alternative solutions. In this case, solution alternatives are pursued in parallel until the decision for an option is made according to the LRM principle (see Section 4.4.2).

It must be mentioned that the Last Responsible Moment should not be confused with procrastination, i.e. the unnecessary postponement of work or decisions. Rather, the Last Responsible Moment is associated with responsible handling and requires some techniques in its implementation. These include:[15]

- Share even incomplete design information in the team.
- Organise direct cooperation between team members.
- Develop a sense of how to manage change.

Basically, it is about finding the *most* responsive moment in which to make the directional decision. In this sense, the following structured but very simple and direct approach can be applied:[16]

1. Do we have a solid, final and watertight basis for decision-making?
   If there is nothing to be gained by postponing, the decision is made now!
2. Do we have to decide now?
   If no other choice is possible, decide!
3. If we decide now (and the answer to 1 is "no"):
   - What are the implicit assumptions?
   - How high is the risk that one of these assumptions turns out to be wrong?
   - What are the consequences of such failed assumptions?
4. If we postpone the decision:
   - What are the immediate and/or unavoidable consequences?
   - What are the possible consequences?

The LRM rule is therefore nothing more than an operational implementation of backward scheduling and its respective benefits described in Section 3.3.2. However, it can also be applied without corresponding explicit planning. In particular, in areas in which too early fundamental decisions, e.g. architecture decisions in software development, correspond to the natural urge for planning security, but also contain the danger of later inflexibility, it should be considered – if the technical possibilities allow it.

4.4.2 Set-based design

Set-based Design (SBD) is a design method that has many applications in the architecture, engineering and construction industries. In the Toyota Production System it has been called

15 see Poppendieck/Poppendieck 2003, pp. 57–61.
16 see Wirfs-Brock 2011.

set-based concurrent engineering, which illustrates how the SBD process differs from traditional point-based design: In set-based design, the project team pursues multiple options simultaneously and eliminates options only as the learning process progresses.[17] Instead of selecting a promising option early on and (only) continuing to work on it, set-based design explores a variety of possible alternatives. The set of possible solutions is gradually narrowed down until it converges to a final solution. During the design process, some options are eliminated due to hard constraints, impossibility or lack of fit. Ideas can also be used to generate more options. At the Last Responsible Moment the decision is made and one of the options is selected.

Known as the second Toyota paradox, for example, it was part of the Toyota Production System early on to consider a wider range of possible solutions, produce more physical models, delay key decisions as long as possible, and still have the fastest and most efficient vehicle development cycle in the industry. Set-based design has also been used in the past on a variety of Lean Construction projects to streamline the design process.[18]

4.4.3   Rolling wave planning

Planning is difficult – especially because it concerns the future!

This variation of a well-known bon mot regarding forecasts, attributed to authors from Karl Valentin to Niels Bohr, describes the dilemma of planning. The more uncertain the future development, the more difficult it is to plan ahead accordingly. The further into the future a plan is supposed to reach, the more uncertain its accuracy generally is – because there is simply more time available for unforeseen and unpredictable events and thus the probability of their occurrence increases. On the other hand, there is the human desire for planning certainty with the need for a planning guideline and the orientation of the project team towards a common plan. If planning is done too early, however, faulty assumptions and a high revision effort are pre-programmed – a prime example of waste.

The management sometimes demands from the project management that detailed planning be presented, even over a rather long time horizon. According to the motto "A good (project) manager must know what is to be done and what is needed for it!" Especially in the planning and allocation of human resources, this increasingly leads to ongoing adjustments and re-planning. For example, it was observed at one client that employees were sometimes allocated to different projects on an hourly basis with a planning horizon of several months. Unfortunately, this over-planning fails to recognise that projects are subject to an inherent, sometimes more and sometimes less pronounced uncertainty that must be dealt with. Nevertheless, in order to take into account the human desire for planning certainty, the need for a planning guideline and the orientation of the project team towards a common plan, planning should be undertaken as a rolling elaboration. This simple rule applies here:

17 see Planview n.d.
18 see Lean Construction Blog n.d.

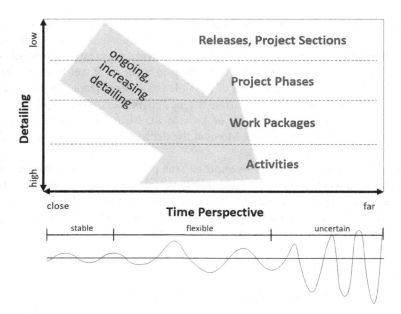

The more distant the planning object is in terms of time, the more coarsely it is planned, and the closer it is, the more detailed this is done.

This means that the granularity of planning becomes finer the closer the planned events, e.g. activities, come. In order to implement this approach, a corresponding planning architecture is required, as shown schematically in Figure 4.6.

At the highest level, the *project sections* should be planned, which often also produce releases of the solution that can be put into production. Such a release comprises, for example, a defined set of functionalities of an IT system. Within the framework of the MuSCoW rule (Must-have, Should-have, Could-have, Won't-have), the first release of a solution may include the must-have items. A project section is defined here as a bundle of phases that produces a result that can be used independently (cf. *programme increment* according to SAFe). This can be, for example, the implementation of a pilot application. A characteristic feature of a project section is that it usually consists of several project phases, e.g. conception, realisation, production preparation, etc. (or also sprints in Scrum-projects) and thus also shows essential characteristics of an independent project. A duration of up to a maximum of nine months is recommended here, which brings advantages such as an increase in flexibility and controllability, not least with regard to budget allocation within a financial year (faster provision of results, increase in flow, etc.).

The planning of project sections is strategic project planning. It is broken down tactically in *project phase planning*. This is where the phases of a project section or a correspondingly dimensioned project are defined. Depending on the process model, these can be the sequential phases of a waterfall-like approach or the cyclical sprints of an iterative approach – or mixed forms of these, as described with the spiral model. Phases are characterised by milestones that define at least the end of the phase in terms of time and content. I have had very good experiences with the introduction of a *golden section* of phases. Loosely based on Leonardo da Vinci (the mathematics behind it is considerably older), the golden section divides the phases into fifths in terms of time and subject matter. After about 2/5, an intermediate milestone should be scheduled, at which fundamental and trend-setting results are available and evaluated for further work. An example of this is the detailed draft of a document to be created, such as a concept, or the basic customising of a standard software solution, etc. The draft should be completed by the end of the project. At this intermediate milestone at the latest, the development of the results should be agreed in advance with the later customers, i.e. the clients of the project phase.

The *work packages* traditionally form the technical-organisational task bundles for processing the project contents. They are typically the leaves in the tree-like work breakdown structure.[19] The planning of work packages should take place in three steps: (1) identification and definition, (2) content structuring (incl. dependencies) and (3) effort calculation and coordination. All this should be done in cyclical cooperation between project management and those responsible for implementation, if necessary in two to three iteration loops. In the sense of Rolling Wave Planning, the following applies: For work packages that are far apart, step 1 is carried out (provisionally) and, if applicable, step 3 is carried out as a rough estimate. This can give the project management and the project team valuable indications of the extent of the tasks ahead. For the upcoming project phase, the work packages must then be concretised through steps 2 and 3. Last but not least, the activities (tasks) to be carried out are identified and planned. A rule of thumb says that the duration of work packages should not exceed one to two weeks; at least this would be helpful. This means that in agile procedures, sprint content and work packages coincide.

Rolling wave planning can thus be summarised as follows:

> High-level planning on target – detailed planning on sight!

A useful field of application for Rolling Wave Planning is human resource planning. A possible and proven design is as follows: Plan the resources:

1. For the next month with the exact person and task – e.g. "Employee John Smith will work on the work package conception with 30% of his capacity in the upcoming month of May".
2. For the following three months with exact roles – e.g. "In the months of June to August, a senior Java developer will be deployed in the web relaunch project at 80%".
3. For the other eight months only roughly, at departmental level – for example, "For the months of September to April, the development department needs web developers amounting to four full-time equivalents."

---

19 see Ottmann et al. 2008, pp. 163–174.

4. Repeat steps 1 to 3 on a rolling wave basis in the appropriate cycle, about once a month, so that the overall planning horizon (in varying degrees of detail) is always twelve months in advance.

With this procedure, the specific employee is always bound for the next four weeks in the agreement between the project management (demand provider) and the department management (demand cover). On the other hand, there is enough time to cover the demand and, if necessary, to adjust resources (e.g. increase staff).

The findings of the *Theory of Constraints* should also be implemented in the procedure. To this end, the resources that are bottlenecks in human resources must be identified and planned with particular attention. For example, if there is only one employee in the company who still knows the old programming language with which the critical running application was created, it is not sufficient to plan here only at the role level. In this case, a longer-term concrete allocation of this person, for instance in the time horizon as in step 2, is recommended. Of course, the second part of bottleneck-oriented planning, namely the removal of the bottleneck, should not be disregarded. This is the only way to eliminate the bottleneck and reduce the risk of failure.

The presented approach to rolling wave resource planning operationalises the classical approach to resource planning, which allocates available capacities to project needs – or in short: "Bring the people to the work". The situation is different in the scenario of the capacity-oriented pull system, as it can be implemented with the project kanban. The pull mechanism turns the planning logic around to "bring the work to the people", i.e. the described steps of rolling resource planning no longer fit. Nevertheless, they can and should be used in a modified form to identify personnel requirements with a corresponding time horizon in order to be able to act in a planning capacity.

With the characterising profile of a project (see Chapter 6), the meaningfulness of project planning, not least of its process planning, becomes clear. The more uncertain the context, the less solid long-term planning is in detail. Here it is important to find the right balance. In any case, Rolling Wave Planning is an adequate approach, because here the degree of uncertainty or instability of the plan is reflected in the time horizons with which detailed planning should be done. The more uncertain the situation, the more it is necessary to plan "on sight". In any case, it will never work without planning. The following insight applies: The plan is nothing – planning is everything! This means that the planning process itself – ideally coordinated with all those involved – creates clarity for those involved in the sense of a discovery process.

4.4.4 Last Planner

From the domain of Lean Construction comes the *Last Planner rule* developed by Ballard at the Lean Construction Institute.[20] This rule alleviates the situation that always arises in a logical, sequential creation of trades that build on each other when the executors of the next trade are not involved in the creation of the input. The predictability of the subsequent work depends on the quality of the immediate planning of the predecessors or on what is handed over to the work

20 see Ballard 2018.

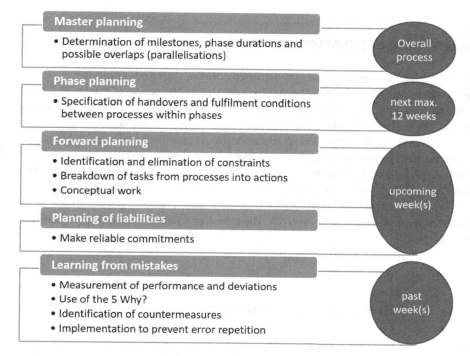

Master planning
- Determination of milestones, phase durations and possible overlaps (parallelisations)

Overall process

Phase planning
- Specification of handovers and fulfilment conditions between processes within phases

next max. 12 weeks

Forward planning
- Identification and elimination of constraints
- Breakdown of tasks from processes into actions
- Conceptual work

upcoming week(s)

Planning of liabilities
- Make reliable commitments

Learning from mistakes
- Measurement of performance and deviations
- Use of the 5 Why?
- Identification of countermeasures
- Implementation to prevent error repetition

past week(s)

FIGURE 4.7   Rolling wave planning with last-planner rule.

as input. As the German saying goes, it's the dogs that bite the hindmost. In order to avoid the "throwing over the fence" of work results and thus this effect, the Last Planner Rule requires that those responsible for the subsequent trades are always involved in the preparation of the predecessor trades and the corresponding planning.

At its core, the Last Planner rule pursues Rolling Wave Planning. For this purpose, Ballard designed the planning and control procedure shown in Figure 4.7.[21]

As shown in Figure 4.7 the level of detail and concretisation becomes more and more specific in the course of the process. The status of planning objects evolves from "should" to "can" and "will" and finally to "is done". "Master plans and phase plans indicate what should be done, when and by whom. The task of forward planning is to prepare planned tasks so that they can be carried out at the planned start. Plans to which all parties are committed and which represent what work will be done are created from work that is ready to be done."[22] "Work packages, where the schedule does not match the actual schedule, are identified in the comparison of 'is done' to 'will be done' and then analysed for possible countermeasures to reduce further impact."[23]

21 after Ballard 2018, p. 124.
22 Ballard 2018, p. 123.
23 Ballard 2018, p. 124.

CLASSIFICATION IN THE MAGIC TRIANGLE

In the classic Magic Triangle of project management, the expected output is first defined as precisely as possible in the project order (scope). During the course of the project, this is considered fixed, and the planning of the effort (time, money) is based on this. In the agile approach, the direction of vision changes: The scope is continuously adjusted on the basis of the basically fixed time (time boxes) and effort (fixed team), so that ultimately a Pareto optimum is created through this *design-to-budget* (see Figure 4.8).

Practices presented, such as the MuSCoW rule and Rolling Wave Planning, now make it possible to combine these views. Finally, the rigid boundaries of the classic Magic Triangle of project delivery are softened. However, this requires a value-oriented reinterpretation of the Magic Triangle, as described in Section 3.1. The practice of Target Value Design in particular offers a solution.

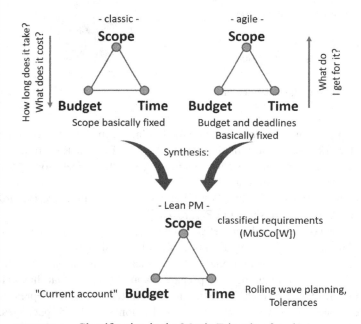

FIGURE 4.8 Classification in the Magic Triangle of project management.

With the practices of goal-oriented, flexible project planning presented in this section, there are also norm strategies with which the "cross of complexity" (see Section 2.3) can be countered. As shown in Figure 4.6 the uncertainty, which makes an important contribution to the complexity of

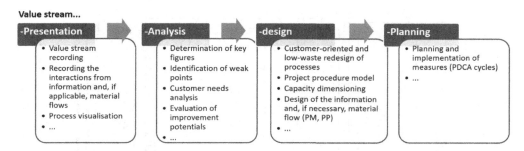

FIGURE 4.9 Procedure of the value stream method.

a task, generally decreases with increasing temporal approach, so that a corresponding planning certainty is gained.

4.5 Process-oriented control

4.5.1 Value stream method

Orientation towards the value stream i.e. the process that generates the output associated with the customer benefit, is a core element of Lean Management. In this respect, it is also applied in Lean PM. It consists of Value Stream Mapping, analysis, design and planning. It involves designing a value stream in which non-value-adding activities are eliminated as much as possible. Figure 4.9 shows the procedure as it can be transferred from the area of production to projects.

In all projects, the information flows must be considered. They are of crucial importance. In particular, they characterise the value creation processes (value streams) of the PM. Not in all, but in many projects, the material flows are of course also essential, for example in construction projects. Especially in the IT and organisation sectors, however, these play a subordinate role. The *Makigami diagram*, which has its origins in the field of Lean Administration, i.e. the representation and optimisation of mainly information and data flows, is therefore recommended for the representation of the process.[24] It is a form of *swimlane representation* of processes, as is also known in the *BPMN method (Business Process Model and Notation)*.[25] Figure 4.10 shows a corresponding canvas developed in the Laboratory for Process and Project Management (PPM Laboratory) at the Technical University of Applied Sciences Central Hesse (GER) for working with the Makigami method.

The focus is on the representation of the process flow with its work steps (sub-processes). In the style of BPMN notation, swimlanes are used which are assigned to the task holders (in Figure 4.10 exemplary O1 to O3) of the respective work steps, so that a change in the swimlane always illustrates an organisational break in the process flow. The task managers of the project area in question, or of the whole project if applicable, are bundled in a *pool* (exemplified by B1

24 cf. Leyendecker 2015.
25 cf. e.g. Freund/Rücker 2014.

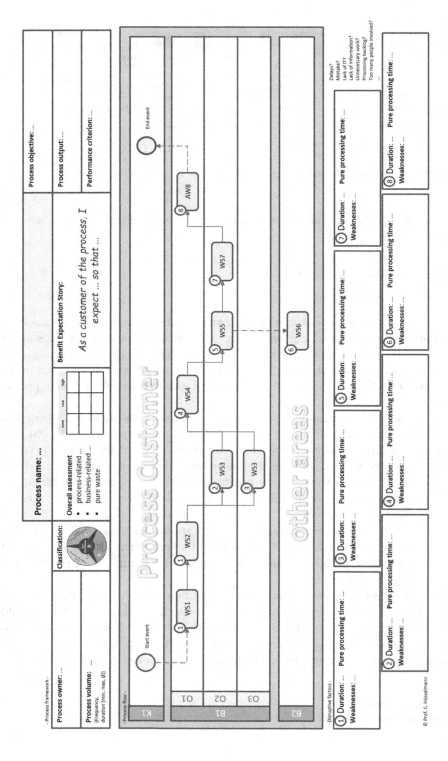

FIGURE 4.10 Process Analysis Canvas.

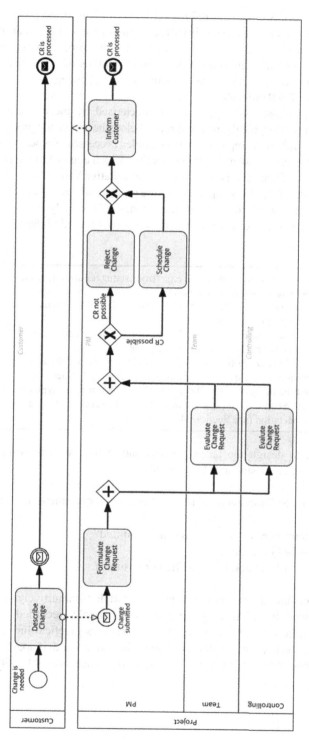

FIGURE 4.11 Simple BPMN diagram using the example of a change request.

in Figure 4.10). The process customer – as the trigger and recipient of the process output – is shown in a separate pool, as are other areas involved in the process. These can be line units in particular, such as purchasing, which are to be involved in the project at certain points but are usually not directly involved in the project discipline (and thus often difficult to control). Figure 4.11 shows a simple example of notation from the context of change request management in projects to illustrate the BPMN method.

Value stream mapping in Lean Management is characterised by the collection of process performance indicators. These are typically placed directly below the associated process step, which experience has shown to be difficult to represent in the complex processes of practice. In the canvas, the corresponding areas are therefore identified by numbering, which marks the connection between the process step and the area of analysis. The focus of the quantitative Value Stream Analysis is the ratio of the value-adding time to the throughput time of the process, which is surveyed or estimated accordingly. The description of weak points of any kind completes the analysis.

In the header of the canvas, there is some information about the administrative and design framework of the process:

Process name:	End-to-end view, if possible, e.g. "from status report to measure".
Process owner:	Who is responsible for the design?
Process volume:	Information on the classification, in particular, of the capacity commitment through the process.
Process objective:	Outcome of the process. Why is the process being carried out, e.g. "Enabling corrective action to continue the project".
Process output:	Tangible result of the process, e.g. artefact, such as a status report
Performance criterion:	Criterion for the quality of the process, e.g. speed of decisions
Evaluation	The type of waste: Not all waste is the same, not every type can be avoided
Benefit Expectation	Story: Sentence template for describing the process customer's benefit expectation (see Sections 4.7.1 and 4.9)

The information is used to design the target process and to assess its relevance. Important guiding questions for the target design are, for example:[26]

- How do we imagine the ideal process if we leave all external requirements out of the equation for the time being?
- Which process steps, interfaces or documents can be omitted?
- Where can tasks be worked on in parallel?
- How can the transfer of information be simplified or ensured?

The Makigami diagram enables a clear process representation that is intuitively understandable. Interfaces become transparent and weak points are systematically recorded. For large and non-sequential processes, however, the representation reaches its practical limits. This can be counteracted by choosing the appropriate delimitation of the process and, if applicable, dividing a top-down view into sub-processes. As a rule of thumb, the scope available within the space

26 see PROMIDIS 2015.

on the developed canvas (see Figure 4.10) can be used – approximately ten process steps or sub-processes per diagram. REFA also recommends the following rules when implementing the value stream method:[27]

- Use paper and pen!
- Team building through integration of all stakeholders!
- View, understand and evaluate at the site of the event (Gemba)!
- Check the data basis and, if in doubt, collect new data!

4.5.2 Project Kanban

Kanban systems were originally used and became known in production. They are used for pull-oriented production control and were first developed by Ohno at Toyota in the Toyota Production System which can be considered the source for the derivation of Lean Production Management and finally Lean Management. In essence, the aim is to establish a self-controlling system in sequential production that makes it possible to avoid overproduction (at individual work stations), to reduce waiting times for material input and thus to keep the flow of production even and low-waste. Simplified, the production in workstation N-1 is triggered by a pull, i.e. a concrete demand message from the following workstation N. This demand message is sent with the help of a pull signal. This demand notification takes place with the help of a material card – Japanese: *Kanban*.

Another central element of Kanban control in production is the standardisation of material flows through uniform containers and, above all, predefined filling quantities of these containers. These are aligned with the average capacities, operation durations and procurement times when designing the system. Production Kanban systems can be designed in different ways, for example by adding intermediate storage facilities as buffers or by linking several preceding work stations. The decisive factor for the functioning and realisation of the desired advantages is that production is sequential, relatively low-variant and in significant quantities. Figure 4.12 shows such a system schematically.

The idea of pull orientation was first applied to the development of software, especially in the operational area of implementing change requests.[28] Here, too, work is usually done sequentially (requirements analysis, conception, realisation, testing, production preparation and finally go-live), often with different workstations, i.e. processors such as analysts, programmers, testers, etc.). The aim was to achieve an even flow of work for the processors (developers) and to avoid or at least reduce overloading and thus inefficiency through the typical pushing of requirements into the work processes. The *IT Kanban system* was born. The value stream described above is depicted in a task board that shows the different phases/work steps of the process in successive columns (see Figure 4.13).

However, the IT-Kanban system differs significantly from the production Kanban system. First of all, the pull mechanism is not demand-driven, but rather capacity-driven. This means

27 see refa.de (German Association for Work Design, Company Organisation and Corporate Development).
28 see Anderson 2003.

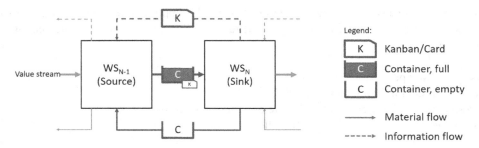

FIGURE 4.12 Scheme of the production Kanban.

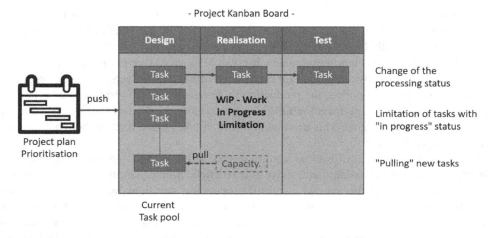

FIGURE 4.13 Schematic task board in the project Kanban.

that the processor of process N pulls an order from the completed tasks of the upstream process N-1 when capacity is free. In order to avoid overloading at this point, *work-in-progress limits* are set for those process steps for which such overloading would be a hindrance – e.g. the simultaneous processing of too many change requests. The WiP limits, measured in workload, story points or simply the number of tasks, thus limit the workload in process step N and are also the benchmark for determining free capacity there.

Another difference between the two systems is the requirement to prioritise the items in a task board column. In *Project Kanban* it is necessary to measure them in terms of their criticality or usefulness. The effort involved is generally not sufficient as a control variable. Prioritisation mechanisms can be of a very simple nature, such as Prio A, B or C, or more differentiated, for example with the WSJF (Weighted Shortest Job First) ratio or the Target Value Design (see Section 4.3.2 or 4.3.3). With Figure 4.13 we extend the usual project Kanban board by a classical project plan as a clock pulse from a higher perspective. The tasks that are due for processing as "next to do" are derived from the plan on a rolling wave basis. A *push-pull mechanism* is created in the sense of a hybrid PM.

In some cases, so-called *fast lanes* are also set up in the task boards, on which highly urgent jobs that arise at short notice are channelled through the system. Depending on the workload with regard to the work-in-progress limit, work that has already been started may have to be put on hold.

The project Kanban system has its origins in requirements management for IT realisations, not least in the area of change request management. As in the production Kanban system, a defined process is mapped and self-regulated in the ongoing (IT) production process. Larger development orders, however, lead to development projects, as here whole teams work on complex solutions that lead to a significantly new solution. The characterising definition of a project should not be repeated here (see Section 1.1), but it is obvious that the design of a project Kanban system requires a further, differentiated design of the IT Kanban system. For example, the project-specific value stream must be taken into account in the design of the task board, as well as the many interdependencies between the tasks. Last but not least, within the framework of the Lean PM concept, a system should be available that is not only applicable to IT projects.

First of all, the value stream of the project or the part of the project to be controlled by the Kanban system must be defined. This forms the basic structure of the task board. Here, of course, the subject of the project plays a decisive role, because it makes a difference, for example, whether an IT application is to be developed or a department is to be reorganised. We have already talked about IT development, so let's take business process optimisation as an example of a reorganisation project. According to Becker et al. the classic process model of a process-oriented reorganisation project consists of the following steps: (1) prepare modelling, (2) develop strategy and regulatory framework, (3) carry out actual modelling and analysis, (4) carry out target modelling and process optimisation, (5) develop process-oriented organisational structure and (6) introduce reorganisation.[29] This provides a possible basic structure for the Kanban board. As agilisation is also taking place in modern business process management – i.e. re-organisations are no longer started with a *big design up front* and end in a late big bang at the go-live of the new organisation – control with the help of a Kanban board offers corresponding potential to make the sequential course of the project more flexible in implementation. The individual technical processes that are the subject of the project scope are the items that are changed, i.e. processed, by the project value stream. Figure 4.14 shows the basic structure of a project Kanban board schematically.

The practice of project management shows that in many cases a task cannot be processed continuously in one piece. For example, an activity triggers an activity of another person or even a team and can only be continued when the latter has reported completion. A typical example is the release of a result by a (team) external authority, such as a line manager. Here it is necessary to remove the tasks that have been started from the work-in-progress calculation so that they do not unnecessarily impede the further flow of work at this point. For this purpose, a (sub-)column of the corresponding process step can be set up ("On Hold/ Waiting") or the corresponding task can be marked, e.g. in colour. If the task can finally be completed, the item is pushed into the further (sub-)column "Done" and is thus available for the next processing step.

29 Becker et al. 2005, pp. 20–22.

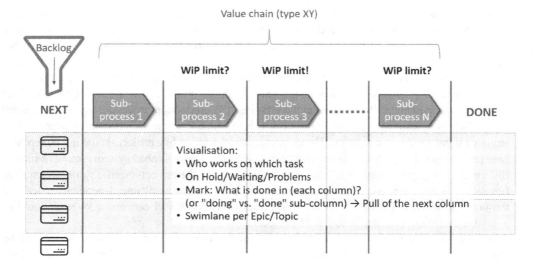

FIGURE 4.14 Value stream oriented design of the project Kanban.

As in the original IT Kanban the project Kanban is filled by a backlog from the left. In a project, this backlog is formed from items in the requirements specification, i.e. from the defined scope of the project. In agile projects, where the boundaries to continuous product development are sometimes blurred (*DevOps*), the backlog is generally more open and subject to more frequent changes. Depending on the project approach, different prioritisation rules are advantageous here – such as the MuSCoW rule for a classic approach or the Weighted-Shortest-Job-First rule in the agile approach. Prioritisation also includes the consideration of urgency. This is generated in projects by given deadlines, such as the go-live of a legally required functionality. In such cases, the latest possible start times for beginning work on a task are calculated by backward scheduling, a form of the pull principle in projects. It is therefore obvious that this deadline requirement is suitable for filling the "Next-to-Do" column of the task board. In the project, operational prioritisation is thus primarily based on the urgency of task completion. Other prioritisations that focus on the benefit of an element are more on the tactical-strategic level of prioritisation (see Section 7.1). Once accepted, elements are subject to operational control in the project.

In the course of a project, further operational to-dos, decision-making requirements or issues continuously arise that cannot be provided for *à priori* in any plan in the world. For the operational control of a project, it is critical that these things are administered in the open points management. In the course of the project, the elements of the project scope and the dynamically created open points together form the backlog of the work to be done (see Figure 4.15).

The items on the *Open Items List* are therefore subject to continuous updating and prioritisation (in every project), and must fit into the systematics of the backlog prioritisation. In my experience, you can read the control-related state of a project from the open points management: "Show me the open points management and I'll tell you the state of the project!"

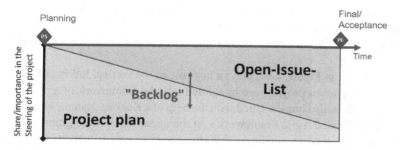

FIGURE 4.15 The Open Items List and the project plan form the backlog.

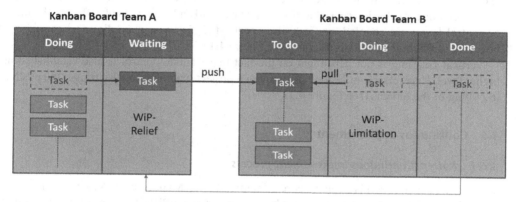

FIGURE 4.16 Horizontally coupled project Kanban boards.

In the course of the project, more and more items of the original, scope-related backlog are processed. On the other hand, new open items are constantly being added that need to be worked through. At the end of the project, the list must be empty, or at least it must not contain any items that prevent acceptance or operation. (A completely empty list of open items is probably utopia and is no longer even required in agile approaches. In any case, the classification of the criticality of the remaining open points is decisive for the future direction of the project). In summary, it can be said that the backlog of a project, whether with or without Kanban control, is composed of the planned and the initially unplanned items that are to be processed.

Another aspect to be considered in the project Kanban is the networking of the Kanban boards of the different teams, e.g. of sub-projects (horizontal coupling), as well as of the different levels – from the portfolio to a programme, if applicable, and the overall project to the individual operative team (vertical coupling, see Figure 4.16).

Work orders that are processed in Team A and that require an activity (additional work) from Team B are pushed into Team B's backlog. There, they are subject to the (pull) control mechanism of team B. Deadline orders are to be given the corresponding urgency. Once the task has been processed by Team B, it moves to the status "completed" and is returned to Team A as available. There it is again subject to the usual clocking-in mechanism, e.g. work-in-progress limitation of the doing column. A variant of horizontal coupling operationalises the passing on of tasks across different teams. The principle is the same as shown before, but several team Kanbans are connected in series. If necessary, the waiting status is omitted if it is not needed to return the results.

The backlog funnel, shown in Figure 4.14, plays a key role in the vertical linking of Kanban boards. It becomes necessary when projects are integrated into the framework of agile project portfolio management, which is also implemented with the help of a Kanban system. An example of this is the SAFe system (Scaled Agile Framework). At the upper level, more granular items, so-called *epics,* are controlled in corresponding value streams. If an epic is activated there, i.e. transferred to the doing column, then it is up to the portfolio management to control this item as a whole or subdivided into finer positions, e.g. user stories, into the Kanban board of the executing team, i.e. to fill the funnel. If necessary, the details are divided among different teams, or a programme level is interposed to divide the tasks. If Kanban boards of different abstraction or control levels are coupled, one can also speak of *flight levels* according to Leopold – i.e. boards at different levels of abstraction (see Figure 4.17).[30]

Vertical coupling is not dependent on the fact that Kanban control is also established at a higher level. Here, classic milestone planning or a project roadmap can also trigger the push, i.e. the filling of the funnels (see Figure 4.13).

4.6 Continuous improvement

4.6.1 Kaizen/Continuous improvement process

The pursuit of perfection is a central core principle of Lean Management. As already mentioned in Section 3.3.3 the term *perfection* must be considered in a differentiated way, especially in the context of projects, since maximum perfection cannot be had for free – and is not necessarily demanded by the customer. However, as a beacon in the sense of efficient, waste-free work, the principle should of course also apply in the project context. In particular, the basically repetitive and standardisable processes of PM itself, e.g. status reporting, are subject to consideration here. But also in the case of recurring work of a similar nature in the technical work of the project, e.g. the software engineering part of an IT development project, the effort of striving for perfection is worthwhile. In both cases, subsequent activities benefit from the preceding experiences and their implemented lessons. Or, to put it the other way round: If, especially in knowledge-intensive project work, one does not draw lessons from experience and use them to shape future activities, this usually results in a significant amount of waste. Inefficient work and mistakes are repeated. Therefore, in projects of any kind, a *continuous learning and improvement process* (CIP) must be established. In this CIP, all those involved in the project – project management, project team and ideally the project customer – should be involved and encouraged to identify improvements and to introduce them into an implementation process.

The CIP approach, which originated in the Toyota Production System, was implemented under the name *Kaizen* which means "change for the better". It is noteworthy that Kaizen

---

30 see Leopold 2017, pp. 26–36.

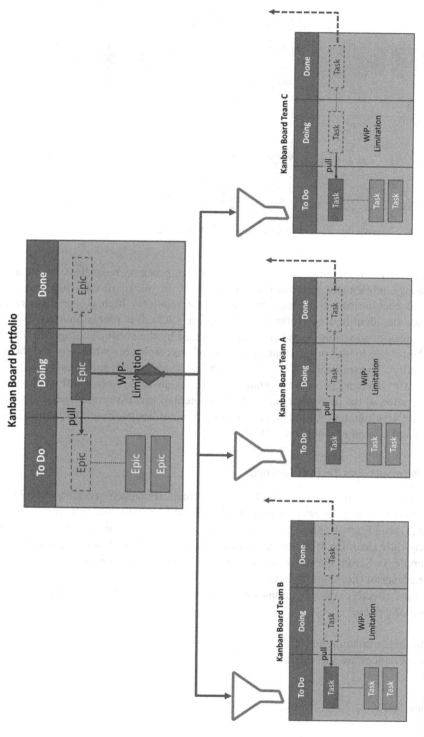

FIGURE 4.17   Vertically coupled project Kanban boards.

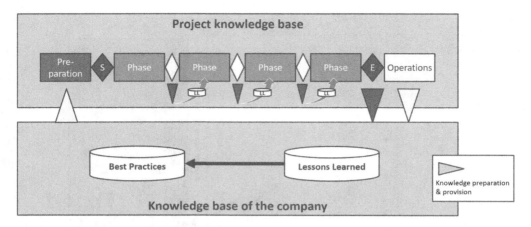

FIGURE 4.18   Knowledge generation, transfer and utilisation.

explicitly calls on all employees involved in the process, regardless of their level, to continuously implement improvement measures.[31] In view of the sociological findings of cultural peculiarities in Japan, which include a relatively high power distance or power acceptance (as identified by Hofstede, see Section 7.1.1), this seems obvious. How else can the bottleneck of superiors be meaningfully replaced by the swarm intelligence of the team – or let's better say by the professional expertise of the employees who carry out the processes operationally? The transfer to the project situation, especially in the classic, plan-driven and hierarchically structured context of *plan & control*, is obvious. Kaizen/CIP explicitly calls for the cooperation of the performers in the improvement of processes, methods and tools.

Kaizen/CIP ultimately results in a continuous cycle of design, execution, critical review and improvement – in other words the *PDCA cycle*.

4.6.2   Retrospectives/Lessons Learned

Ideally, learning from experience in the sense of the CIP happens continuously. However, in order to ensure the implementation of the CIP, this should be systematically planned for at phase transitions (see Figure 4.18). In classical methods, e.g. PRINCE2 this has been firmly anchored for a long time (learning from experience) and agile procedures place a special focus on this. For example, in Scrum the *retrospectives are the procedures explicitly intended for* this at the end of each sprint (i.e. short phases).

The most effective way of gaining Lessons Learned is to hold a Lessons Learned workshop at the end of a phase:

- The timing is favourable.
- The experiences are still fresh.

31 see Imai 2002.

- Essential experience for the next phase and new projects can be gathered, evaluated and secured.
- Tools used (templates, checklists, etc.) and processes of cooperation should be optimised.

Helpful questions are at hand:

- What was particularly good and should be adopted for follow-up projects?
- What changes were there to the plan and what were the causes?
- Which checklists, templates etc. need to be modified?
- What should be done differently or maintained at all costs in the coming phase?
- What can be adopted for a similar new project or must be done differently there?

As the questions show, the Lessons Learned are not only about the disruptive factors that made the work unnecessarily difficult, but also about the success factors. It is all too easy for these to be overlooked and, if applicable, not consistently implemented in the subsequent phases. The results of the Lessons Learned workshop should be documented and the resulting measures taken. The following tried and tested form of documentation can be used:

ID:	Unique identifier of the position
Subject area:	Serves to cluster the position in order to bundle measures and/or implementation responsibility wh ere appropriate, e.g. communication
Factor:	Naming of the identified influencing factor, e.g. poor user involvement in the design
Short description:	Explanation, background information on the position
Category:	Success factor (S) or disruptive factor (D)
Influence:	Assessment of the significance of the position, such as high – medium – low, to derive a prioritisation of measures.
Measure:	Designate a point of action to address the influencing factor, e.g. conduct regular reviews with the users.
Brief description of measure:	Explanation, background information on the measure
Remark:	Other comments on the explanation

The definition creates new to-dos, which are entered in the *project backlog* (*open items list*, see also Figure 4.15 and Figure 5.10)! There they are supplemented with further administrative elements, such as due date or responsibility, so that the tracking of execution is made possible.

By transferring the identified and evaluated lessons into concrete measures, it is avoided that Lessons Learned only become lessons *lost*, as Kloss has described from many years of experience.[32] He identifies five statuses for dealing with Lessons Learned that can be found in practice in the project context: *Lessons Demand Detected, Documented, Presented, Accepted* and finally *Converted*. In many cases, only the "presented" stage is encountered in practice, i.e.

---

32 see Kloss 2019, p. 211.

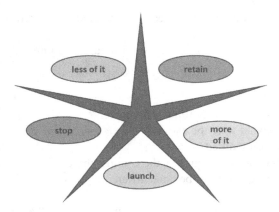

FIGURE 4.19 Starfish method for retrospectives.

the lessons are not consistently implemented and the findings are then lost – to put it in Lean PM terms: The effort of developing them is wasted. It is only at the "converted" stage that the Lessons Learned bring the benefits associated with their name.

The time intervals of these Lessons Learned workshops is usually based on the length of the project phases. The shorter a project lasts, the more frequently the Lessons Learned must be carried out if they are to be of any utility for the rest of the project. However, a practical rule of thumb is that individual project phases should not extend over more than three months, even in longer projects (otherwise it would make sense to split them up). This determines the latest possible cycle for Lessons Learned. With Scrum the phases, i.e. the sprints, last for example two to four weeks, which results in the respective cycle of the retrospectives.

The more frequently workshops are held, i.e. the less time there is between two consecutive workshops, the more stringent and light-footed they should be, otherwise there is a risk of a fatigue effect – quote from a Scrum team member: "What are we supposed to discuss in the 25th retrospective?" For example, a maximum duration of 1.5 to 2.5 hours is required for retrospectives in Scrum – depending on the sprint length. To avoid waste, the meeting can also be shorter, but they should not be omitted completely. If the project phase and the intervals are longer, such a Lessons Learned workshop can also last half a day to a whole day. As a rule, this should be anything but a waste of time and capacity, which is something you often have to convince the project seniorities of. The subsequent benefit will prove it.

A lightweight method for conducting Retrospectives/Lessons Learned workshops is the *Starfish method,* which is described in Figure 4.19.

The five fields of this "starfish" speak for themselves and are very suitable as a visualisation in a team-wide workshop. In combination with the Lessons Learned documentation, the Starfish exercise can also be used as an introduction to teamwork. The "Influence" field of the documentation is then designed in a more differentiated way according to the five fields of the Starfish.

No matter which tool is used in the end: In the follow-up workshop, the documentation of the previous workshop should be retrieved and the state of implementation and any new findings compared with it.

4.7 Further practices

In the following, a few more practices are listed and briefly explained that are noteworthy and useful in connection with the implementation of Lean PM and that can be transferred directly from PM or from Lean Management. There is no claim to completeness here in view of the almost inexhaustible variety of methods and tools.

4.7.1 Voice of the Customer and User Story

Voice of the Customer (VoC) is a method from the field of Lean Management/Six Sigma and is used in quality management (improvement processes) as well as in the development process (new development). Voice of the Customer collects the expectations of the target group (customers) regarding the process or the product. Influencing factors are then derived from the customers' expectations, which are objictified in figures and summarised in measurable and evaluable variables. In this way, statements can be made about the degree of fulfilment and the number of errors (number of processes that do not meet the customer requirements, etc.). At the end of Voice of the Customer, there is a weighted list of customer requirements that is backed up with measurable variables.

User stories, on the other hand, are a method from agile software development. A user story presents a specific functionality of the product that can be implemented in one iteration. It represents the type of communication between the product owner and the programmer. User stories can be formulated vaguely at the beginning and become increasingly detailed in the course of the process – all in all, they can be characterised as requirements formulated in a lightweight manner, the implementation of which requires further specification with suitable (software) engineering methods, e.g. the class diagram. User stories require a fixed sentence template in a format like

As [ROLE], I would like to [WISH] so that [USE].

The combination of both methods brings together their advantages. The starting point is the collection of the opinions and wishes of the potential customers or clients. These are then fitted into the sentence template of the user story and, as far as possible, assigned to a domain. Measurable key figures are derived from this assignment, which can be used to read off the progress of the domain and the individual aspects. A target value is set for these indicators. This defines the point at which the corresponding aspect maps to the desires derived from Voice of the Customer, and thus this measurable target value forms the acceptance criterion for the user story (see Figure 4.20).

The simple example shows that one sentence of the Voice of the Customer can lead to several user stories and these must be assigned to different domains. Here, prioritisation is important in order to classify the influence of the domain on the customer's request. In our example, server performance has a much higher influence on waiting times. Thus, the server performance has to be prioritised via an adapted script language.

The combination of both methods developed in the PPM Laboratory of the University of Applied Sciences Central Hesse (GER) thus offers potential to improve the result. However, this

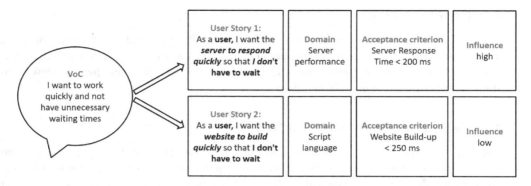

FIGURE 4.20 Template of the combined Voice-of-the-Customer/User-Story method in the example.

also requires the will and the possibility to carefully and systematically capture the Voice of the Customer. Since the aim here is to include the opinion of the end customer as representatively as possible, this is associated with corresponding effort.

4.7.2 Expense calculation

There is regularly a great deal of uncertainty in effort estimation, which is simply due to the novelty for those involved and the risks of many projects. Waste in the form of estimation errors to be corrected (renegotiations), misguided resource allocation or simply repeated planning is the result. The following techniques have proven helpful in tackling this task.

4.7.2.1 Three-Point Estimate

In order to take the inherent uncertainty of the estimate into account in projects the probable effort (R = realistic estimate), the minimum effort (O = most optimistic case) and the maximum effort (P = most pessimistic case) are estimated for each estimation object, e.g. work package. Finally, planning is based on an expected value (E), which can be calculated as a weighted average from the three estimates mentioned. In the empirically derived *PERT variant (Program Evaluation and Review Technique),* this is done with the formula $E = (P + (4 \times R) + O)/6$. Last but not least, based on the deviations of the values P, R and O from each other, a very good estimate of the existing uncertainty can be made and corresponding risk contingencies can be planned for.

4.7.2.2 Effort Driver-based Estimation

In order to determine the effort of a process, its *effort drivers* can be determined and quantified, comparable to activity-based costing. For example, for the work package "Programming evaluations", the number of evaluations is the decisive indicator, sensibly divided according to complexity levels (simple, medium, high or similar). The effort drivers can be materialistic, as in the example, but also time-related (the effort for PM as a so-called *job stream* depends not least on the duration of the project) or at first glance very indirect. In yacht building, the rough cost calculation per running metre of ship can serve as an example. The decisive factor is that there is a sufficiently strong correlation between the effort driver and the real effort. The estimation based on the effort driver can be combined very well with the three-point method.

4.7.2.3 Planning Poker

Planning Poker has become known as a practice from Agile (Scrum). In a playful procedure, efforts are estimated by the team. The procedure is an on-site variant of the Delphi method, as the accepted planned value is found in a multi-stage procedure with feedback from the participants. As in a poker game, the estimates of each individual participant are first determined covertly, then revealed and finally the differences are discussed in order to gain insights for the next round of estimation. Typically, in Planning Poker, the effort is rated in the form of story points in a Fibonacci series (1, 2, 3, 5, 8, 13, 21, etc.), which abstract from the actual time required and serve only as an indication and subsequent control variable.

Finally, my own experience has shown that the combination of methods often yields the best result in hindsight, by combining different top-down and bottom-up methods and, if there is fundamental agreement, accepting it as a realistic estimate.

4.7.3 Defect avoidance

Having to correct errors in processing is certainly one of the most serious forms of waste. Producing errors always means extra work and often even unsuitability of the result, even danger to life and limb. The following selected methods can contribute to defect prevention in projects.

4.7.3.1 Poka Yoke

In general, it takes more effort to correct mistakes than to avoid them in the first place. *Poka Yoke* stands for avoiding or completely preventing accidental errors or defects that arise from carelessness. For this purpose, physical conditions of products (e.g. the shape of a plug), result checks by measurements (e.g. the weight of a produced part) and the error-averse design of the process flow (e.g. the next step can only be carried out after the previous step has been confirmed with a signal) are applied. The *hard* Poka Yoke prevents a certain error from being made at all, the *soft* Poka Yoke reduces the probability of occurrence or detects made errors before further processing.[33] The use of checklists is an example of a soft Poka Yoke.

APPLIED POKA YOKE IN THE EXAMPLE

Delays in the project process are no exception. Often these delays are due to the complexity of the tasks to be carried out in a division of labour – work package A waits for the input that is

---

33 see e.g. Bertagnolli 2018, p. 125 ff.

to come from work package B. In a large project with own participation (ERP implementation in a large conglomerate), it has proven useful to specify for each work package in its definition from which other work packages service is expected. In contrast to a complete central network of causal dependencies, this created a bilateral view with qualified information. With the help of a suitable IT system and an appropriate organisation, an early warning system could be set up in this way, in which, for example, the delivering team is reminded (pull) two weeks before the due date and can give feedback. This reduced the "active waiting" that can unfortunately be observed in (large) projects, which leads to delays.

4.7.3.2 Nudging

Another interesting method that has proven itself in practice many times over, which can be assigned to the soft Poka Yoke, but does not originate in Lean Management, is *Nudging*. The Nudging method uses simple, mostly visual aids to lead people unconsciously to desired actions – without prohibitions, commandments or economic incentives. Human behaviour is thus influenced in a predictable way.

A striking example to explain the method of Nudging is an application to increase safety at work in the company through a human-sized "shadow man" on the wall of the staircase, who uses the handrail and wears the company name on his chest: He reminds the employee of the right action at the time of the decision (of climbing the stairs), serves as a role model, appeals to the employee's self-image with the logo, appears sympathetic through his green colour and signals like a traffic light: "Now you can go!"[34] Other examples often found on the shop floor of production companies are conspicuous red markings on the floor that indicate danger zones to be avoided. Similar examples are often found in the implementation of the 5S method, where tool storage areas are visually marked according to their shape.

4.7.3.3 Failure Mode and Effects Analysis

In order to find meaningful Poka-Yoke practices, the first step is to identify the typical PM defects that are based on carelessness and chance. *Failure Mode and Effects Analysis* (FMEA) is a general method that can be applied to projects, especially in quality management in product development and production. It serves to minimise risks through the early detection and prevention of potential errors and their effects on product functions.[35] Process-related FMEA focuses on the establishment of flawless processes for the manufacture of components and products. The following points are to be dealt with:[36]

---

34 see Mields/Kuzcynski 2020, p. 3.
35 see Romeike 2018, p. 85 ff.
36 see Kuster et al. 2011, p. 421.

- Listing of all conceivable faulty components (risks).
- Indication of the possible consequences of the problem.
- Formulation of the causes of a problem and corrective action.
- Probability of occurrence (O).
- Effect of the error (E = carrying distance).
- Probability of detection in the company (D).

The values O, E and D should be set in a precisely defined rating scale (usually a Likert scale). The product of the three quantities results in the *risk priority number* (RPN), which describes the level of a risk: $RPZ = O \times E \times D$. The RPN generates a prioritisation; the root cause analysis is used not least to identify preventive measures in the sense of the Poka Yoke.

4.7.4 Cause identification

4.7.4.1 (Typed) Ishikawa diagram

A well-known tool for cause-and-effect analysis is the *Ishikawa diagram* (ID), also called a *fishbone diagram* because of its appearance. It looks for the causes of an identified problem and classifies them into the clusters of *man, machine, method, material, environment, measurability and management* (7M) or variants thereof. The Ishikawa diagram can thus be used, for example, in the context of FMEA analysis to determine or limit possible causes of defects. The graphic representation is reminiscent of a fish structure – hence the alternative name of the method developed by the Japanese scientist Kaoru Ishikawa.

In my own experience, the challenge in practice often is that, on the one hand, several problems are identified during a process analysis that need to be remedied, and on the other hand, the causes of these problems cannot be clearly assigned, but may cause several problematic effects at the same time. Thus, there is usually no 1-to-N relation between problems and causes, but rather an M-to-N relation. To depict this with the conventional Ishikawa diagram, where one diagram is created per problem, M diagrams would have to be created – with the attendant likely redundancy of multiple causes. This creates waste and in practice is observed to be a barrier to the use of the method.

In the work of the PPM Laboratory of the University of Applied Sciences Central Hesse (GER), a multi-dimensional Ishikawa diagram was developed that we call *typed* (see Figure 4.21).

As shown in Figure 4.21 the Typed Ishikawa diagram makes it possible to list several problems in one representation. As with the usual Ishikawa diagram, the causes sorted according to the 7M clusters are also assigned to the respective problems. To clarify the respective cause-effect relationship, the corresponding relation is typed, i.e. marked with a coloured identifier.

In one project, for example, the low data quality in the goods receipt process was caused by "no timely posting" – both were consequently marked with the same identifier "x". However, the lack of timely posting also led to delays in the further course of the process ("y"). In order to document this, the aforementioned cause was given the second link to problem "y", etc. By

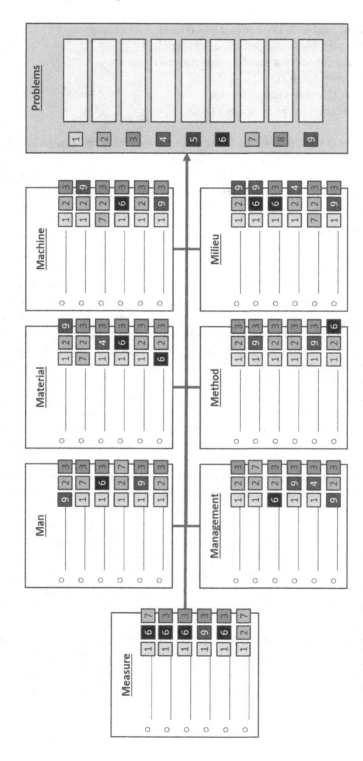

FIGURE 4.21 Typed Ishikawa diagram.

simply counting the documented links, if applicable adding a problem weighting, the identified causes can now be prioritised and remedied in bundled measures.

4.7.4.2 5W questioning technique

In problem solving, it is important to identify the root cause of a problem and to correct it. It is often found that the immediate cause is not the fundamental cause because it has itself been triggered by another cause. It is therefore important not to work on (intermediate) symptoms, but to get to the bottom of the actual causes. In the *5W questioning technique,* the respective reason is asked until there is no more answer to this question – one feels reminded of the questioning technique of a small, curious child. Experience shows (although not always) that after asking five times, one will have found the causal reason for one's problem, hence the name 5W question technique (5 times why?). Ohno himself describes the following example:[37]

EXAMPLE OF A 5W QUESTION TECHNIQUE

1. Why is the machine no longer running?
   Because the fuse has blown due to overload.

2. Why was the machine overloaded?
   Because the bearings were not lubricated.

3. Why were the bearings not sufficiently lubricated?
   Because the oil pump was not pumping.

4. Why didn't the oil pump pump?
   Because the shaft is knocked out.

5. Why was the shaft knocked out?
   Because there was no sieve and metal chips got into the machine.

The genuine cause has been found and can be eliminated: Sieve installation instead of (repeated) fuse replacement.

In projects, problems are particularly revealed in written or verbal status reports. The 5W questioning technique is therefore useful in analysing status information and subsequently deriving measures. This also prevents the same problems from recurring in the further course of the project.

---

37 see Ohno 2013, p. 51.

4.7.5 Facilitating work

4.7.5.1 Visual Management

A key concept of Lean Management is the creation of transparency about the state of the system. For most people, transparency and the cognitive perception of non-trivial issues is enhanced by a visual representation. The famous traffic light example makes this clear: Worldwide and regardless of the level of knowledge, people recognise that a red traffic light means "Stop! Danger!" and a green traffic light means "Go! All is well!". It is no coincidence that the traffic light metaphor is often used in operational processes, be it to report the status of a machine, to release further processing or to indicate the status of a project. Visualisations are prominently used in Lean Management as an action principle called *Visual Management*. Imai, a pioneer of Kaizen defines Visual Management as a practice for the transparent preparation of information and instruction regarding individual elements of work with the aim of enabling individual employees to improve their productivity.[38] By visualising facts – including quantitative ones that can be evaluated by key figures – they become easier to understand and those involved can recognise deviations and correlations more quickly. The visualisation in Figure 4.22, based on Bertagnolli, illustrates the difference between ambiguity and transparency in a striking way – simply by the way the information is presented.[39]

Can you see in the left-hand illustration whether all types of points, a, b and c are present the same number of times, or which point might be missing? The appealing visualisation on the right certainly makes the answer much easier for everyone. Wherever possible, this effect should be transferred to everyday project work. Typical areas of application are risk and progress presentations, cost analyses, status information, resource load etc. Many of the Lean methods use visualisation intrinsically, e.g. Kanban, soft Poka Yoke, Ishikawa diagram or Value Stream Mapping.

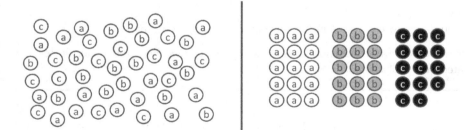

FIGURE 4.22   Effect of a targeted visualisation.

38 see Imai 2002, p. 386.
39 after Bertagnolli 2018, p. 334.

4.7.6 5S method

The *5S method* also uses visualisation as a formative element. The aim of this method, is to improve the working environment, reduce unnecessary searching, increase productivity and follow clear rules of conduct at the level of the individual workplace. On the basis of a corresponding standardisation, continuous improvement can be established and, in particular, it enables employees to work efficiently in cooperation and in the possible change of processing (workplaces). The 5S method includes (in this order):

1. Sort out! (Seiri)	Material and tools that are no longer needed, i.e. work equipment, are removed.
2. Clear up! (Seiton)	Required work equipment is kept within easy reach, shelves and storage areas are (visually) marked with the corresponding item.
3. Keep the workplace clean! (Seiso)	A clean workplace increases safety, quality and also motivation at work.
4. Make an order to rule! (Seiketsu)	Elimination of doubts in the execution of work through transparent, visible instructions; thereby achieving routine.
5. Keep all points and constantly improve! (Shitsuke)	Motivation to comply with the 5S and their continuous improvement; rewarding and making improvements visible.

The 5S method, as it is easy to see, originated in the field of production. Nevertheless, it is easy to identify areas of administrative work to which the 5S can be applied analogously. In addition to the classic design of the office workplace (filing, registration, desk, etc.), the administration of electronic documents should also be mentioned. Let him who has never searched for a file or accidentally processed the wrong one cast the first stone! In practice, there is an enormous need to catch up, which, according to my own estimates, is only temporarily covered by modern collaboration platforms such as wikis (keyword: Full-text search). Structured document storage, naming conventions, versioning, deletion of files that are no longer needed, etc. are effective elements of implementing 5S in this area – and thus in project business.

4.7.7 Complementary practices

The analysis in the PPM Laboratory at the University of Applied Sciences Central Hesse (GER) made it possible to identify a number of other practices that tend to originate in the area of production but could also be usefully employed in projects of any kind. These include, for example, the *8D* and *A3 reports* as variants for status reporting, the *Andon board*, the *shop floor board* and the *Heijunka board* for visual support of process control, the *5-cycle check* and *PDCA cycle* for standardised process control and, last but not least, *bottleneck identification*, which is also known from the network planning technique in classical PM. The *indicator-based evaluation* of the state of the processes is also an important principle of action in Lean Management, which is also classically relevant for the control of project processes. Its specific characteristics are discussed in Section 7.2.2 in the context of the introduction of Lean PM. The list of practices can be continued almost indefinitely. A description of further methods and tools would go beyond the scope of this context.[40]

4.8 Agile or classic? – the Agilometer

The balance between stability and flexibility has to be found.

Agile elements are included in the solution spectrum of the design of a hybrid PM system in that iterative or incremental procedures are recommended in certain constellations in the goal-oriented design of PM – this is a core element of agile methods (see Chapter 6). Modern PM is increasingly about the question of balancing *agile* and *traditional* approaches.[41] These two poles of PM are not only about the procedural model in the sense of the project process, but also in particular about the questions of leadership and general corporate culture. Irrespective of the original characteristics of a project, the corresponding environmental factors are also relevant for the choice of approach: A bureaucratic-hierarchical organisation will not lead an agile approach to success. A chaotic-creative organisation will basically have difficulties with adherence to plans (see Section 7.1.1).

For some years, therefore, researchers and practitioners have been concerned with the question of which characteristic features of a project and its environment can explicitly answer the question of whether a plan-driven hierarchical (*traditional*) or a rule-based iterative (*agile*) approach is possible and useful. Boehm and Turner in particular have developed a foundation here, which has been used and supplemented by authors such as Špundak, Timinger, Feldmüller et al.[42] The criteria known through this include:

Environmental conditions:	Project-internal conditions:
Corporate culture	Stability of the requirements
Qualification level of the team members	Criticality of product safety
Alignment of the organisation in projects	Project team size
Team distribution	Complexity of the project subject
	Product adaptivity

At this point, *product adaptivity* refers to the property of the products to be developed and further developed in increments, which Feldmüller et al. measure with reference to the costs of changes as well as the duration of development.

THE STACEY MATRIX AS A CLASSIFICATION TOOL?

The so-called Stacey matrix is often used as a decision-making aid to clarify which approach – agile or plan-driven – is best for a specific project. Numerous online articles bear witness to

---

40 see e.g. Bicheno/Holweg 2016.
41 see Timinger 2017; Seel/Timinger 2017.
42 see Seel/Timinger 2017; Brehm et al. 2017; Blust 2019.

this. In the current version of the Stacey matrix, the dimensions of uncertainty regarding the requirements (let's call them UR) versus uncertainty regarding the procedure (UP) are usually listed. The calculation of complexity then supposedly results from the formula complexity = UR x UP.

However, the Stacey matrix is rather a diagram that makes it possible to classify existing problems according to the two dimensions mentioned above. In his publication, Stacey himself uses the dimensions (degree of) agreement versus (degree of) certainty. With this classification, Stacey derives the adequate complex management method that should be used, e.g. Brainstorming and Dialectical Inquiry, if the problem is characterised by a relatively low level of agreement as well as a relatively high level of certainty. Stacey does not mention a classification of the degree of complexity!

With the explanations on complexity (see Section 2.3), it also becomes clear that complexity cannot be spanned with the criteria of uncertainty regarding the requirements and the procedure alone. In particular, aspects of the diversity of the elements of the system under consideration (complicacy) as well as emergence and dynamics must be taken into account.

The use of the modified Stacey matrix may be helpful in practice in some cases, but it is partly misleading. Who can carry out a highly complex project in a purely agile way? And it was not conceived this way by Stacey himself. It no longer appears in more recent editions of his much-cited basic work![43]

When analysing the remarks of Boehm and Turner in particular, one finds that a distinction must be made between the possibility and the meaningfulness of the same when choosing an approach. Feldmüller also picks up on this by examining the utility of using agile elements.[44] Ultimately, one should consider whether a methodological approach is possible, whether it brings benefits and should therefore be used, or whether this is even required, i.e. necessary, due to the constellation (see Figure 4.23).

The use of a method becomes possible when the professional-technical as well as the organisational prerequisites are given. Diebold and Simon describe the latter in terms of the application of the behavioural model of human action as *wanting, being able* and *being allowed*.[45] These three aspects must be in harmony in order for a project to be possible, such as the application of agility (see Figure 4.24).

The Wanting, for example, assesses the management's requirement to practise agile procedures. The Being-allowed includes compliance requirements that demand, for example, the separation of creators and testers of a solution, which makes close cooperation impossible. For example, there are legal regulations that prevent agile elements, e.g. DevOps, in the area of human resources. There, the separation of development and operation is mandatory for data protection reasons. Finally, the Being-able describes, for example, the organisational prerequisites that must be in place, such as the physical proximity of the project teams including customers for an agile approach. These examples can of course be extended in many ways.

---

43 cf. Habermann n.d.
44 see Feldmüller 2018.
45 see Diebold/Simon 2019.

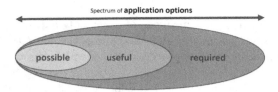

FIGURE 4.23   Spectrum of deployment options.

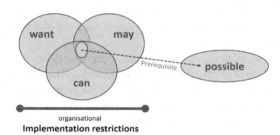

FIGURE 4.24   Organisational implementation restrictions.

As shown in Figure 4.23, the next higher level of stringency always requires the previous one. If the use of one method is required, then the use of the opposite method is essentially impossible. In the evaluation and detailed design, this ultimately leads to a specific catalogue of criteria and its application to the question "Agile or traditional approach?" (see Figure 4.25).

This *Agilometer* combines the already known influencing factors with a number of additional factors from empirical findings. They can be determined with the following guiding questions from the professional-tactical, the organisational and the corporate cultural context:

Professional-tactical features:

- *Security with regard to the procedure or the technology*
  Is the underlying technology and the associated procedure known in the project work and is there any experience in this respect?
- *Stability of the requirements*
  Is it to be expected that the technical requirements will be subject to strong, possibly repeated changes at least for the duration of the project?
- *Certainty about the goal*
  Can the goal be clearly formulated, ideally operationalised in concrete terms, and is it shared by all stakeholders?
- *Structural complexity/Complicacy of the project subject*
  Is the project subject of a simple structural nature or does it consist of a multitude of interconnected components?
- *Release of increments/product adaptivity*
  Is the product to be developed insensitive to late changes or are these expensive and difficult to implement?

	Criterion	Occurance Level						Approach
		low			high			
		possible	useful	obligatory	possible	useful	obligatory	
Professional-tactical features	Safety with regard to the procedure or technology	x	x	x	x			agile
					x	x		traditional
	Stability of the requirements	x	x	x	x			agile
					x	x		traditional
	Security with regard to of the goal			(x)^l	x			agile
					x	x		traditional
	Structural complexity/complexity of the project object	x			x			agile
		x			x	x		traditional
	Release of increments/Product adaptivity	x			x	x		agile
		x	x		x			traditional
	Requirement for project support documentation	x			x			agile
		x			x	x		traditional
	Criticality of product safety/Hazard potential	x						agile
		x			x	x	x	traditional
Organisational features	Quick wins required	x			x	x		agile
		x			x			traditional
	Desired/prescribed level of detail in planning	x						agile
		x			x	x	x	traditional
	Intensive exchange within the team, with users/customers possible				x	x		agile
		x	x	x	x			traditional
	Availability and commitment of client & product owner				x	x		agile
		x	x	x	x			traditional
	Success through interdisciplinary cooperation	x			x	x		agile
		x			x			traditional
	Experience and skills for agile working/level of qualification				x	x		agile
		x	x	x	x			traditional
	Project team size	x	x		x			agile
		x			x	x		traditional
Corporate cultural features	Flat hierarchy given	x			x			agile
		x	x		x			traditional
	Target system at group/project level in place	x			x	x		agile
		x	x		x			traditional
	Flexible company processes in place				x			agile
		x	x	x	x			traditional
	Agile affinity of the management/Alignment of the organisation				x	x	x	agile
		x	x	x				traditional
	Employees prefer many degrees of freedom				x	x	x	agile
		x	x	x				traditional
	High level of trust between project client and project service provider				x	x		agile
		x	x	x	x			traditional
		possible	useful	obligatory	possible	useful	obligatory	Option
		can-	should-	must-have	can-	should-	must-have	

FIGURE 4.25 The Agilometer – shaping the deployment options of agile and traditional approaches.

- *Requirement for accompanying project documentation*
  Do the circumstances require extensive accompanying documentation with regard to the production of the results, e.g. for the complete tracing of the process?
- *Criticality of product safety/hazard potential*
  Is the finished product used in a safety-critical manner and thus has a high hazard potential, especially for users and the environment?

Note on target safety (see (x)[i] in Figure 4.25): If the goal cannot be defined, the use of an agile approach is not sufficient, because these approaches also need a goal (e.g. product vision) – even if not as specific as the plan-driven ones. A procedure that has become known as *effectuation* can help here.[46] Starting from a problem to be solved, solutions are developed opportunistically on the basis of existing possibilities and a defined, limited budget for measures. In this sense, effectuation expands the solution space from the plan-driven approach beyond the agile approach into the area of dealing with even greater uncertainty.

DEALING WITH UNCERTAINTIES IN THE CONCEPTION

In conceptual design, one often faces uncertainties about the feasibility or cost of the concept. It is worth making efforts to address these uncertainties. Since it is better to find out as soon as possible if a certain aspect of the concept cannot work, a small investment upfront at this point can save a lot of time and money on the rest of the project.

For this purpose, the following sample for handling and mitigating conceptual uncertainties is taken from the example of a project for the introduction of robotics in production:[47]

Unanswered question:	*Will the camera be able to find the input parts reliably?*
Hypothesis:	Yes
Conviction:	Low
Is it absolutely necessary to have the right answer?	Critical
How can we validate the hypothesis?	Testing with robot, camera and parts (method for testing and measuring results)
Estimated time and costs for validation:	The supplier can validate within one working day; we already have the equipment.

---

46 see Heinen-Konschak/Brendle 2017.
47 after Bouchard 2017, p. 141.

Organisational features:

- *Quick wins required*
  Do results, and if applicable partial results, need to be available quickly and ready for use?
- *Desired/prescribed level of planning detail*
  Do stakeholders, e.g. the senior management, require detailed planning, e.g. to approve budgets, etc.?
- *Intensive exchange within the team, with users/customers possible*
  Are the organisational prerequisites in place for the team to be able to exchange information closely with each other, but also with the subsequent users of the result?
- *Availability and commitment of the client and product owner*
  Are the client and/or the product owner highly available during the development of the project outcome and able to influence it?
- *Success through interdisciplinary cooperation*
  Does the project promise more success if the team brings together a variety of experts who can thus contribute different perspectives to the solution?
- *Experience and skills for agile working/level of qualification*
  Are the intended team members highly qualified so that they are able to deliver the required results independently and responsibly – and not also teamwork?
- *Project team size*
  Does the project team include many employees (> 100) or few (< 10)?

Corporate cultural features:

- *Flat hierarchy given*
  Are there many hierarchical levels in the company and is a hierarchical decision-making culture practised?
- *Target system in place at group/project level?*
  Do employees have target systems that (only) reward their own performance or are group targets (also) in place?
- *Flexible company processes in place*
  Are bureaucratic processes established in the company, e.g. in the area of procurement, or can a non-bureaucratic and flexible approach be taken here?
- *Agile affinity of the management/alignment of the organisation*
  Does the management support and live an agile mind set and is the organisation already aligned for the implementation of agile projects?
- *Employees prefer many degrees of freedom*
  Are project staff particularly productive when they are given many degrees of freedom or do they prefer clear work instructions to guide them?
- *Is there a high level of trust between the project client and the project service provider?*
  Is there a good relationship of trust between the project client and the (external or internal) service provider that is also forgiving of mistakes and strives for joint solutions – or is a contractual arrangement rather more advantageous?

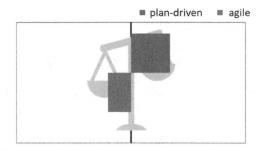
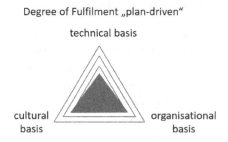

FIGURE 4.26 Qualitative, visual analysis.

The *Agilometer* was applied, for example, at an automotive supplier in the course of selecting and introducing a Manufacturing Execution System. The result of the analysis was a hybrid PM approach in which some elements of agile working with a low implementation hurdle for the company (due to the existing knowledge in the area of Lean Production) were gradually integrated into the project work and combined with a classic, plan-based approach. These included, for example, the product vision, the product backlog, a project Kanban board, the implementation of increments on the one hand and a phase and milestone plan, a steering committee or explicit risk management on the other.[48]

In general, the practical application of the Agilometer does not provide a clear recommendation that is supported by all criteria in the same way. In any case, due to the diversity of the criteria and the organisations considered, this cannot be assumed. It is methodically possible to calculate an "agility score" by assigning a value to each characteristic and adding them up to a score – if applicable even weighted among each other – similar to a value benefit analysis.[49] However, this procedure appears to be overengineered and a qualitative, visually supported evaluation (such as in Figure 4.26) is preferred,[50] and serves as an aid for the solution that is ultimately the responsibility of the PM.

In particular, the overall scoring does not take into account the absolute criteria required, as shown in Figure 4.25 (classified as *required*).

4.9 The PM Value Stream Analysis

4.9.1 The identification of value streams

The application of Lean Management includes the identification of the value streams of the domain as well as their design according to Lean principles. Figure 4.27 shows the central high-level end-to-end processes (value chains) of PM.

With the help of the processes defined in the reference model of the Unified Project Management Framework the value chains can be orchestrated. As shown in Figure 4.10 in

48 see Wilhelm 2019.
49 see Timinger et al. 2018, p. 171.
50 cf. Hüsselmann et al. 2019.

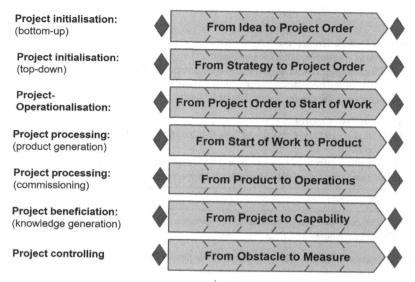

Project initialisation: (bottom-up)	From Idea to Project Order
Project initialisation: (top-down)	From Strategy to Project Order
Project-Operationalisation:	From Project Order to Start of Work
Project processing: (product generation)	From Start of Work to Product
Project processing: (commissioning)	From Product to Operations
Project beneficiation: (knowledge generation)	From Project to Capability
Project controlling	From Obstacle to Measure

FIGURE 4.27   The PM's central high-level end-to-end processes.

Section 4.5 an important element for applying the Lean PM concept is thus in place. Further, customers and their expectations and related wastes should be identified and Lean principles and practices applied. Finally, performance indicators should be defined to quantify the state and development of the processes.

Figure 4.27 can be further subdivided into sub-models that deal with the corresponding PM sub-disciplines (see Unified Project Management Framework). A complete list would go beyond the scope of this book. Therefore, two processes in which the Lean PM application is played out will be dealt with in the following as examples. In general, it must be said that these can only be examples, because ultimately the analysis of weaknesses must and can only be carried out on the concrete application in the company. Nevertheless, the following examples should provide some suggestions. Risk management was selected from the strategic PM area and knowledge management from the area of enabler processes. Other areas, e.g. quality management or stakeholder management, are very similar in structure, so that an analogy is often obvious.

4.9.2   Lean Project Risk Management

Processes	
Processes according to UPMF:[51]	Identify strategic risks [1.3.1]
	Implement risk management [2.3.1]
	Analyse risks [2.3.2]
	Plan risk measures [2.3.3.]
	Monitor risk development [4.3.1]
	Implement risk measures [4.3.2]

51 incl. the UPMF process numbers.

Value chains:

RM1 "From Project Idea to Risk Management System"

O → [Identify strategic risks] → [Implement risk management] → O

The aim is to establish a functioning risk management system (responsibilities, processes, methods) for the project on the basis of the necessities identified by the strategic risk analysis of the project. If available, appropriate company standards can or should be used.

RM2 "From Uncertainty to Risk Control"

O → [Identify strategic risks] → [Analyse risks] → [Plan risk measures] → [Implement risk measures] → [Monitor risk development] → O

The aim is to avoid uncontrolled effects from hazards and to realise opportunities as they arise ... through the operational management of risks (hazards and opportunities).

Customers:

Process customers: RM1: Project Manager, Corporate Risk Management
RM2: Project Manager, Client

Benefit Expectation (sentence template: *"As a customer of the process of the process, I expect ... so that ..."*):

RM1: BES1: "As a project manager, I expect that effective risk management is practised in the project to avert potential threats to the success of the project, so that the project can be carried out within the framework conditions (deadlines, budget, deliverables) set by the project assignment."

BES2: "As a project manager, I expect the project's risk management to be designed to be responsive so that unforeseen events can be managed with as little negative impact as possible."

BES3: "As a project manager, I expect that opportunities that arise in the project are also pursued beyond the framework set by the project mandate, so that the potential of possible outcome improvements (deadlines, budget, deliverables) is realised."

BES4: "As a corporate risk manager, I expect the project to establish a functioning risk management system so that the risks relevant to the entire company can be systematically managed."

RM2: BES1: "As a project manager, I expect the project to systematically identify and assess uncertain, possible events that may have an impact on the course of the project, so that targeted risk management measures can be initiated."

BES2: "As a project manager, I expect the defined risk-related measures to be implemented and assessed in terms of their impact and development to ensure the associated effects to achieve the project's objectives."

BES3: "As a client, I expect that undesired events that have an impact on the project outcome are actively managed to minimise the impact on the business case of the project."

BES4: "As a client, I expect opportunities to be actively exploited so that the business case of the project improves where appropriate.

Domain-specific waste:

Pure waste:
- #The topic of risks is raised too frequently, for example with every status update.
- #Risks are addressed too infrequently or only pro forma
- #Treatment of all (too many) possible risks
- #Too many people dealing with the issue of risk management
- #Too detailed assessment of the identified risks
- #Incorrect valuation system
- #Focusing on the wrong risks, ignoring the real risks
- #Inadequate tool for managing the risks
- #Identifying risks without following up with adequate measures, etc.

Process-related waste:
- #Documentation of the project's risk management system
- #Failure to observe undetermined, unforeseeable events, etc.

Business-related waste:
- #Reporting on risk management in the direction of corporate risk management → no waste from the perspective of corporate RM as process customer
- #Compliance with company-wide risk management standards → no waste from the perspective of corporate RM as process customer, etc.

Application of lean core principles:

Flow:
The flow object is the individual risk. After the identification of a risk, a stringent implementation of the measure derived for it should take place depending on the assessment. The measures should be kept small and flexible (even if synergies are to be considered).

Pull:
Risks should be dealt with proactively – if they have already occurred, then it is a concrete crisis. But the prioritisation of measures leads to the well-known priority- and capacity- or work-in-progress-limit-oriented processing, if a Kanban system is established in the project. It is recommended to use a risk radar to identify possible new risks or changed assessments. The radar then "pulls" the risks into the risk management system.

Perfection:
Perfection is achieved when, on the one hand, the project's risk management system functions efficiently (i.e. with little waste) and effectively (i.e. with effective measures against the right risks). On the other hand, the intention of the overarching corporate risk management is to be achieved.

Performance indicators:

Effort driver: Number of actively managed risks
Number and scope of risk prevention measures
Number and scope of crisis response measures
Frequency of the risk-measure review
Number of persons in charge of risk management
Extent and level of detail of the risk management system documentation, etc.

Key figures:	Overall risk assessment of the project
	Number of risks identified
	Number of actively managed risks
	Budget of the risk measures
	Frequency of the risk-measure review
	Change in overall risk assessment (over time) etc.

Application of lean practices:

Specific:	FMEA	Shopfloor Board
	CIP	Risk radar
	Ishikawa-diagram	Risk register, etc.
	Plan-Do-Check-Act cycle	
General:	Standardisation	RACI matrix (Responsible –
	Value Stream Analysis, Makigami	Accountable – Consulted –
		Informed) etc.

4.9.3 Lean Project Knowledge Management

Processes

Processes according to UPMF:	Implement configuration management [2.10.1].
	Administer project system configuration [3.10.1]
	Control knowledge and documents [4.10.1].
	Prepare final report [5.10.1].
	Implement knowledge and document management [2.10.2].
	Ensure project progress documentation [4.10.2].
	Describe PM system (project) [2.10.3].
	Transfer project results into operation [5.2.3].

Value chains:

KM1 "From Project Assignment to Knowledge Management System"

O → *[Implement configuration management]* → *[Implement knowledge and document management]* → *[Describe PM system (project)]* →

The goal is to establish a functioning knowledge and configuration management (responsibilities, processes, tools) for the project. In doing so, appropriate company standards can or should be used, if they exist.

Establishing company-wide knowledge management is not the goal.

KM2 "From Project Outcome to Knowledge Utilisation"

O → *[Ensure project progress documentation]* → *[Control knowledge and documents]* → *[Transfer project results into operation]* → *[Prepare final report]* → O

Goals are ...

... increasing the productivity of the project team by systematically enabling the use of existing knowledge (data, information, skills) with reference to results produced in the project as well as knowledge available in the company (good/best practices, standards).

... the enrichment of the company-wide knowledge management in the technical domain of the project as well as the PM.

Customers

Process customers:	KM1: Project Manager, Team
	KM2: Project Manager, Team, User Representative, Corporate Knowledge Management, Operations Management

Benefit Expectation (sentence template: "*As a customer of the process of the process, I expect ... so that ...*"):

KM1:

BES1: "As a project manager, I expect effective knowledge and configuration management to be practised in the project to increase the productivity of the team."

BES2: "As a project manager, I expect the configuration management for the project's products to meet the requirements of a smooth operational transition."

BES3: "As a team member, I expect that project and company knowledge is available to me and that I can draw on existing knowledge so that I can carry out my project tasks efficiently and effectively." (Efficient: "not reinventing the wheel", using best/good practices. Effective: Produce good quality).

KM2:

BES1: "As a project manager, I expect knowledge relevant to the project to be systematically documented and managed to facilitate or even enable the team's work."

BES2: "As a team member, I expect that the knowledge needed for the project and in the course of the project is easily and purposefully available so that I can carry out my project tasks efficiently and effectively."

BES3: "As a corporate knowledge manager, I expect the findings (knowledge gain) of the project to be made available beyond the project so that other projects can also benefit from it."

BES4: "As Operations Manager, I expect the project's knowledge of the outputs generated and to be used operationally to be available in operations so that operations run smoothly, and reactive measures can be targeted in the event of disruptions."

Domain-specific waste:

Pure waste:

• Processing (analysis, documentation) of knowledge that is subsequently no longer used
• Redundant development of knowledge
• Loss of knowledge
• Reduced quality of results due to non-use of best/good practices
• Reduced productivity due to non-use of best/good practices
• Time-consuming search for existing knowledge etc.

Customers

Process-related waste:
- Documentation of the project's knowledge and configuration management system
- Analysis, preparation and documentation of the knowledge generated in the project for the project team and the operation
- Introduction of a knowledge management tool
- Establishment of a Knowledge Manager
- Opportunity costs due to the capacity needed to prepare knowledge, etc.

Business-related waste:
- Analysis, preparation and documentation of the knowledge generated in the project for future projects etc.

Application of lean core principles:

Flow:
The flow object is the individual knowledge element. Once the knowledge element has been created, it should be prepared and made available in a timely manner. The processing of knowledge should be appropriately timed to achieve an even load. Knowledge elements that trigger an immediate consequence of action (e.g. decisions) should be communicated promptly and purposefully (push).

Pull:
Knowledge elements that need to be used when required should be made available in such a way that the user can access the content in a targeted way at the required time. Knowledge elements that are to be incorporated into the company-wide knowledge base should be "pulled out" of the project by Corporate Knowledge Management.

Perfection:
Perfection is achieved when the project's knowledge management system functions efficiently and effectively.

Performance indicators:

Effort driver:	Number of employees involved in the project
	Number of relevant artefacts
	Number of knowledge-generating projects etc.
Key figures:	KPIs of the knowledge management domain that value intangible knowledge capital (Intangible Assets)
	Frequency of Lessons Learned
	Budget for the preparation of knowledge (capacitive, monetary) etc.

Application of lean practices:

Specific:	CIP	Tool use (DMS, KM, configuration database)
	Plan-Do-Check-Act cycle	Result logging
	Visual Management	Versioning etc.
	Error collection list	
	Quality control chart	
General:	Standardisation etc.	

5

USE CASES OF LEAN PROJECT MANAGEMENT

After reading this chapter, you will know ...

- some exemplary applications of Lean PM in terms of creating a situational hybrid project approach
- using the example of a large ERP implementation and reorganisation project and a large IT infrastructure and software project.

How can the Lean PM approach be used in the design of concrete projects? In addition to a systematic, preliminary adaptation of the project-specific PM system based on the basic characteristics of the project (see Chapter 6), it is of course also possible and sensible to carry out such an adjustment on the basis of insights or situational requirements identified during the course of the project. In this case, the – often unforeseen – development of the project requires that the PM reactively finds methods and ways to ensure the further successful course of the project.

In the following, I describe the case-by-case, partly reactive, partly proactive use of Lean PM practices in practice on the basis of two of my own large IT and organisational projects. One was the nationwide introduction of a personnel management system and the other was the countrywide introduction of a control centre system.[1]

5.1 Case study 1 – Personnel management system

The "PVS project" provided a prime example of the unproductiveness of waste in PM processes – to the point of acutely endangering the entire handling of the project worth millions. For example, in response to the increasingly difficult situation in the project, the original project manager

---

1 see also Hemmann/Hüsselmann 2006; or Arndt 2016; Steeger 2014; BPUG 2014.

DOI: 10.4324/9781003435402-6

demanded an increased pace in reporting. In order to cope with the amount of information, an assistant was even hired to process the now weekly incoming status reports from a multitude of work packages and sub-projects. The result was that the situation of the project only got worse, as neither was the PM able to convert the flood of information into control measures, nor could the reporters concentrate on their actual productive work. When taking over the project management, the first thing to do was to abolish the high-frequency, paper-based reporting system, to design the communication structures in a targeted and personal way, to revise the distribution of tasks in and with the entire team, and thus to put the project on the road to success. But one thing after the other.

5.1.1 The project at a glance

A few years ago, I was involved in a large SAP implementation project in a federal ministry as an operational project manager on the contractor side. The project involved the introduction of a personnel management system based on SAP HCM (Human Capital Management). In addition to the Federal Ministry as the highest federal authority, the project also included the associated departments in the area with several hundred offices throughout Germany. Some of the departments were independent higher federal authorities with the associated organisational sovereignty. In total, there were about 70 such higher and intermediate authorities. The department employed almost 30,000 staff. The contractor side was formed by a consortium of a consulting firm, a system integrator and the software manufacturer.

The project ran for a total of about six years, with all its ups and downs. A particular difficulty was the first-time introduction of the commercial standard software SAP in a public authority structure of this size. Particularly in the area of human capital management, the civil service differs significantly from the private sector due to its separate wage and salary laws. Also, the cameralistics – only marginally relevant in the project – represent a significant difference to the double-entry bookkeeping and cost-performance accounting of the private economy. The aim of the project was to replace the heterogeneous legacy system landscape with the current release of SAP HCM. This would enable decentralised personnel management with centralised data storage. The project was embedded in a strategic initiative to *Modernise administrative tasks through Business process orientation and IT usage* in the department.[2]

CHALLENGES OF THE PROJECT

- Planning security over a relatively long period
- Precise availability of staff from the specialist department who are not in the core team
- Sufficient availability of staff in the core team
- Variety of requirements
- Variety of tasks
- Large number of employees in the project
- Need to increase efficiency through learning curve
- Time constraints, deadlines, etc.

---

2 see Bürmann/Hüsselmann 2008.

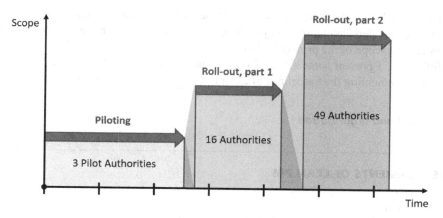

FIGURE 5.1   PVS project strategy.

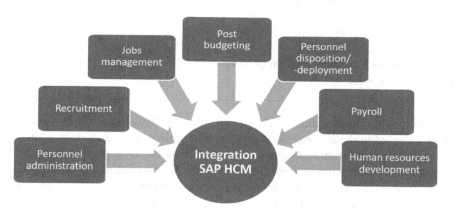

FIGURE 5.2   PVS project – process scope.

The project was carried out in several project sections (see Figure 5.1).

In a preliminary section, a feasibility study including prototype realisation, business process analysis, data conception and profitability calculation according to federal standards was carried out. In the second project section (piloting), SAP was realised and implemented as a personnel management system with legacy data transfer and realisation of interfaces. Three representative authorities were piloted in terms of their characteristics (size, self-sufficiency, pay scale, etc.), which were also put into production.

In the third section of the project, which was divided into two parts, the solution was rolled out in 16 and 49 other authorities of the department. A total of 140 application administrators, 1,300 specialist users and 900 other users (executives, company representatives) with more than 27,000 personnel cases were put into production at approximately 200 locations. In the course of the project, there were repeated successful external audits within the framework of an ISO-9000 certification (contractor) and by the Federal Audit Office (Federal Ministry).

A particular challenge was the necessary and desired optimisation, harmonisation and standardisation of the personnel, post and job management processes of the entire department. The process-related scope of the project is shown in Figure 5.2.[3]

In the following, I present some selected management elements of the project that can be recognised as implementing the guidelines of Lean PM – including a negative example.

5.1.2   Personnel and organisation

APPLIED ELEMENTS OF LEAN PM

- Project mission statement
- Problem solving before contract fulfilment
- De-bureaucratisation
- Dissolving the interruption of the process flow due to the administrative principle of co-signing
- Tandem formation (Pair Working)
- Participation and delegation
- Core team members: Full-time and disciplinarily assigned
- "Small steering committee"

In the course of the first project phase (realisation & piloting), massive frictions arose in the project. These even led to the concrete consideration of a project termination with contractual reversal. In addition to the technical complexity, which was certainly present and probably underestimated by all participants, cultural factors could be identified as reasons for this.

On the one hand, on the contractor side, the partners of the consortium had different ideas about how to handle such a project. This was due to the respective cultural backgrounds of the companies. For example, a medium-sized, young consulting firm has different ideas about a project than a DAX software company. Again, other framework factors come into play in the case of a large system integration company whose employees are employed under collective agreements, partly with a civil service background. If the employees of this consortial partnership were to form a joint project core team, the internal change management effort could not be big enough. In our example, the one-week team-building measure at the beginning of the project was not enough to bridge the differences in corporate culture – even though it was very effective on a personal level, which had a positive effect in the further course of the project.

On the other hand, the public authority staff on the client side had little or no project experience, especially not in a project of this size. The actions of an administrative authority follow different rules than those required for the project business. This includes in particular the assumption and *delegation* of decision-making responsibility in and for the project. Decision-making processes in an administration are usually characterised by co-signing procedures. In these, decision papers are prepared by a technically leading body, which are then circulated

---

3 Further information on the project can be found in Hüsselmann/Hemmann 2006; Bürmann/Hüsselmann 2008.

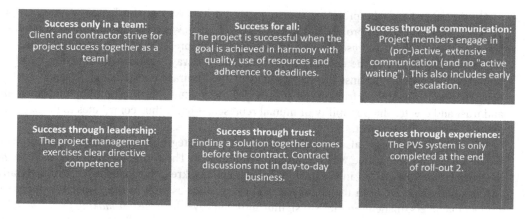

FIGURE 5.3   Project mission statement of the PVS project.

in the administration for co-signature in an often broadly sequential manner. Back-loops and unexpected delays are programmed in here – especially since those involved in this way are fundamentally not involved in the project discipline and goal setting. It is easy to imagine that this leads to massive disruptions in a project that is driven by time pressure and in which the further progress of subsequent work often depends on directional decisions.

In order to get a grip on this situation, among other things, a *project mission statement* was jointly developed that was accepted and welcomed by all participants. Some of its postulates seem like excerpts from a (well-known) manifesto (see Figure 5.3).

This mission statement expresses some fundamental success factors for the project. For example, it recognises that such a complex project, which represents a major financial gamble for all parties involved, can only be carried out successfully if it is always kept in mind that win-win decisions have to be made. Win-lose situations lead to one party being disadvantaged and therefore unlikely to (or even able to) participate in a purposeful way in the further course. From a contractual point of view, the project was a contract for product delivery and the work had to be delivered at the agreed fixed price. In this respect, it seems particularly noteworthy that the postulate *"Joint solution finding takes precedence over contract fulfilment!"* was formulated. The reference to the Agile Manifesto with its postulate "We value collaboration with the customer more than contract negotiation." is obvious.[4] Similarly, the approach here can be seen as applying the idea of a *relational contract* – albeit not in the sense of *a priori* design, but only reactively.

With the aspects of delegation of responsibility, communication, solution orientation, etc., some essential elements of Lean PM can be recognised. But at first glance, there are also rather untypical elements, such as the clear directive competence of the project management. At second glance, however, it becomes clear that a technical and process-related directive competence has a clear efficiency-increasing effect, especially in a project that involves the introduction of standard software. This aspect should be seen not least against the background of the principle of co-signing in the administration described above and makes it possible to maintain the *flow of the project process* at this point.

4 see agilemanifesto.org n.d.

On the other hand, the project was also organised in a very goal-oriented way right from the start. The project, sub-project and work package leaders were always appointed in *tandem,* consisting of a client representative and a contractor representative. This ensured that the requirement criteria and the solution approaches were always worked on together, which is remarkable especially against the background of the contract constellation. Even if such a constellation is not always frictionless, it provides for an atmosphere of mutual understanding and trust and creates the possibility of mutual representation. It thus contributes to the stability and resilience of the project.

The tandem approach also came into play at the strategic steering level of the project, where a steering committee was traditionally established. The formation of a so-called *small steering committee* (KLK, in German "Kleiner Lenkungskreis"), consisting of commercially responsible managers from the contractor and client sides as well as the respective operative project managers, made it possible to significantly improve the *decision-making flow. The KLK* met at close intervals or on an ad hoc basis and, as an important, albeit not sole, part of the steering committee, was given the authority to make decisions of overriding importance at short notice so that the course of the project was not impeded by waiting for the next regular steering committee. In case of doubt, it was accepted with a manageable risk that decisions made by the KLK would not be confirmed by the overall steering committee and that corrections would have to be made. The latter remained a de facto theoretical construct, so that the procedure proved itself completely.

Finally, it is worth mentioning that at least the core team members of the client and parts of the contractor were assigned to the project on a *full-time basis.* This avoids unproductive additional workload due to multitasking and relieves them of dispositive tasks in the interaction of project and day-to-day business. It can also be identified as a success factor that the client's project staff were also assigned to the project management in disciplinary terms for the several years of the project, so that an interface with line management was also eliminated here. (However, this proved to be disadvantageous when, towards the end of the project, the team members were left to their own devices to clarify their subsequent employment due to the lack of a personnel development concept).

5.1.3 Procedures and project flow

APPLIED ELEMENTS OF LEAN PM

- Timeboxing
- Standardised Lessons Learned after each phase or timebox
- Learning curve through iterative-incremental project strategy and lessons converted
- Structuring, standardisation and timing of the *jours fixes*
- Planning architecture and rolling wave planning
- Avoiding multitasking by temporary suspension
- Levelling the workload
- Delegation, operational self-organisation

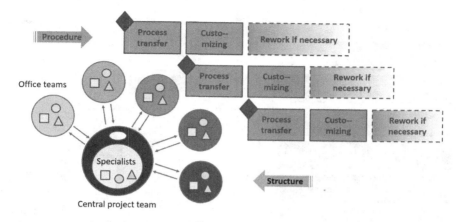

FIGURE 5.4   Iterative-incremental approach to roll-out.

The procedure in the course of the project has some elements of an *iterative-incremental* nature, also characterised as agile procedures. These include, in particular, **timeboxing** (as an implementation of the Lean principle of *pulsing*), which was used in the roll-out phase to integrate the many different organisations in the department (see Figure 5.4).

In the process, regionally organised teams consisting of a fixed core team as well as decentrally integrated, temporary employees in so-called *office teams*, depending on the department, worked on integrating the organisation under consideration into the central, previously developed solution. In a human capital management project, for example, the local company agreements or regional tender channels for staff positions etc. are included. Finally, in this project, the successive process-related expansion of the solution took place in several iterations up to data transfer, training and going live.

In order to achieve maximum planning security for the teams on site for months to come, six-week project phases (sprints) were carried out according to a fixed procedural scheme (see Figure 5.5).

These sprints were carried out repeatedly with different technical process focuses, whereby the solution was processed centrally at the end of each sprint and transferred to the SAP system. Open issues were identified, evaluated and prioritised and finally transferred to the follow-up phase if required (see Figure 5.4 and Section 5.1.4). Finally, a short concluding phase of consolidated system adjustments took place before the subsequent production start-up preparation.

A major advantage of the timeboxing approach proved to be the planning reliability for the decentralised site staff who were only to be temporarily involved in the project. These could be planned on the basis of the postulate "time takes precedence over completeness" using the project schedule (see Figure 5.5 on the right), and they could plan their presence for weeks and months – including substitution arrangements, etc.

The regular central working meetings of the (core members of the) office teams with the "guardians of the solution", i.e. the central teams (week 1 *know-how transfer*), which took place every three weeks, also enabled closely timed feedback on the improvement of the procedure

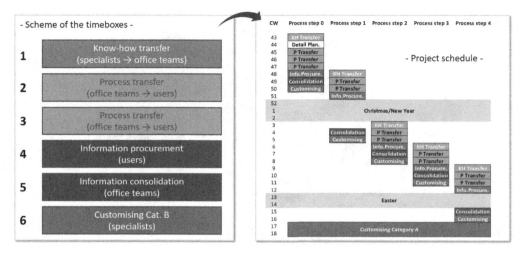

FIGURE 5.5 Office integration in sprints.

ID	Subject Area	Factor	Brief description	Category	Impa
1	Project planning	Detailed planning rollout	Systematic, early and precise implementation planning of the rollout (especially the sub-phase of authority integration) brings security and stability to the project process. Corresponding communication. Particularly in the official environment, this approach meets the clients' need for security.	S	high
2	Contract management	Decoupling Contractual iss	The decoupling of personnel responsibility from operational project management means that the operational project manager is relatively unburdened in day-to-day business. PL can act relatively unencumbered in day-to-day business.	S	high
3	Requirements management	Requirements specification	The requirements specification is insufficient in terms of both process and technical aspects. Process-related: direct contact between customer and developer in a game of ping-pong. Functional: no specification (e.g. template) for specification.	D	medium
4	Project administration	File administration	Web-based file access including document management (versioning) enables and facilitates cooperation, especially between different locations. Locations	S	low
5	Team building	Win-win situation		S	high

FIGURE 5.6 Lessons Learned Register.

and solution (comparable to the retrospective or review in Scrum), so that a constant *Lessons Learned* process was established.

The procedure described here also refers to the project phase of the roll-out, which is characterised by significant procedural repeatability, which suggested this procedure. The identification of Lessons Learned was, however, also an important component of each phase transition in the course of the project. As shown in Figure 5.1, short phases of consolidation were introduced at the transitions from one project phase to the next. These could be used to catch up on any work that had been left undone, but especially to collect Lessons Learned in a timely manner. For this purpose, a retrospective workshop was conducted with the help of a *standardised template* (see Figure 5.6). Last but not least, these consolidation phases also served the personal recovery of the team members, which makes an important contribution to *levelling the workload* in a project lasting several years.

Lessons Learned are part of modern PM and are indispensable in an iterative or incremental approach. The trick is not to let them degenerate into a mutual apportionment of blame, but to always look ahead in a solution-oriented way. The project mission statement helps here.

The project implementation strategy (as shown in Figure 5.1) required, not least, an essential learning curve for all involved. The project phases were scheduled for approximately the same length of time. However, first three, then 16 and finally 49 authorities had to be mapped in the system, which required a significant increase in efficiency. This is only possible through a *systematic CIP*, which is based on practicable knowledge and configuration management as well as the cyclical analysis of disruptive factors, but also of success factors.

Timeboxing was also applied on a small scale, as it is described in the Scrum procedures. This refers to the *timing of work meetings*, which were also subject to a strict time regiment – without preventing a necessary professional discourse. An example of this is the PM *jours fixes*, which were held every two weeks.

In the *jours fixes*, there is a regular exchange on the status and procedure of the project. Due to the size of the project with its sub-projects and the tandem appointments, a large number of people take part in these working meetings. It is therefore almost impossible that for the duration of the meeting all persons are constantly involved in all topics discussed. A significant *waste of working time* is the result (and patience is required) – if the working meetings are not clearly structured. In the PVS project, the *jours fixes* were therefore divided in such a way that the status of the sub-projects was first presented in a fixed time slot and central information was distributed. In this way, all participants were informed about the current developments in the project. Furthermore, this time window was used to identify, document, prioritise and allocate discussion and clarification needs.

In the second part of the working meeting, the previously identified topics were then worked on professionally in targeted personnel constellations – as far as this was directly possible. This made the *jours fixes* very efficient and also effective, as it avoided participants having to spend time on topics that were irrelevant to them, and many points could be clarified directly during the meeting.

If we turn our gaze back to the overall course of the project, it is still worth mentioning that a systematic *planning architecture* was helpful, on the basis of which *rolling wave planning* was carried out. Figure 5.7 gives a visual impression of this.

It becomes clear that starting from a strategic project planning of the project sections, a centrally administered milestone plan was kept. While the time scope of the first level encompassed the entire project, the immediately upcoming project section, which in turn could also include (partial) phases, was planned on this second level. The milestone plan was created in the consolidation phase on the occasion of the project section transitions.

On the basis of the central milestone planning, which was the responsibility of the operational project management, decentralised, operational implementation plans within the time horizon of the coming project phase could now be drawn up in cooperation with sub-project managers and work package managers. Here, the milestones were less relevant than the work processes and detailed scheduling necessary to develop the project results. In this way, the *delegation principle* could be implemented and the team members planned – on the basis of the central high-level specifications – the implementation of their work. The horizontal and vertical consolidation of these plans was the task of the overall project management.

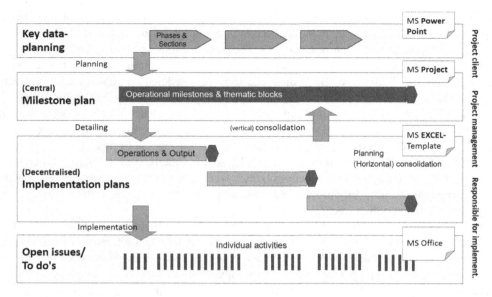

FIGURE 5.7 PVS planning architecture.

In the day-to-day business of the project, the activities of the implementation level were supplemented by identified *open issues* (e.g. in the context of the *jours fixes*), which are referred to in Section 5.1.4 in detail.

In the course of the project, which lasted several years, changes in the framework conditions had to be dealt with several times. A major influence was the fundamental change in the relevant collective agreement, which is very challenging for an ongoing human capital management project. The project was faced with the question of how to integrate this change into the project process.

A *value benefit analysis of* various alternatives led to the decision that the new tariff system should take the form of the insertion (PVS 2.0) of a new project section. Consequently, the project section of roll-out 1 was interrupted and the insertion was processed. In this way, the complexity of such a complication could be kept manageable and ***harmful multitasking*** could be avoided by focusing on the respective tasks in sequential order.

5.1.4 Scope and task management

APPLIED ELEMENTS OF LEAN PM

- Rolling backlog (task prioritisation, open issue management)
- Process orientation (value stream, flow)
- [Waste: Number of reports too high].

The scope of the project in terms of content and organisation was defined within the framework of a fixed-price contract (incl. requirements and functional specifications). Due to this contractual commitment, a flexible handling of the general scope was not possible without further ado. Changes that were nevertheless necessary, such as the aforementioned PVS 2.0 insertion, were organised accordingly within the framework of a systematic, formal *change request management*. This referred (exclusively) to change requests with contractual relevance, e.g. release plan of the system, or to elements defined by the steering committee, e.g. milestone dates. Project elements below this level were handled within the PM team or in the "small steering committee" KLK and documented in a binding manner via decision protocols.

These changes in detail also included consequences from factual situations that arose during the timed roll-outs. Not in all cases a timebox completely delivers the expected result. Unexpected circumstances, organisational complications, lack of information, etc. influence the processing. Thus, in the transition from one timebox to the next, it must be determined which open points are to be prioritised and how, where rework is necessary and how this is to be planned into the next timebox. This ultimately results in a ***rolling wave task planning*** from timebox to timebox, which is only fixed for the current timebox. Often, tasks that were originally scheduled could be transferred from one organisation to another, for example by analogy, and thus removed from the backlog. The technical basis for this was the hierarchical ***process model of personnel administration*** created in the first project phase. This conceptual basis served both as a content-related technical basis and as an organisational-technical framework (see Figure 5.8).

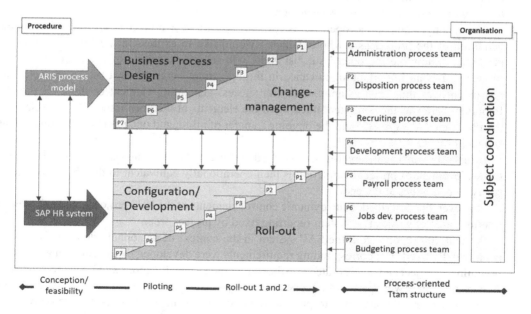

FIGURE 5.8   Process orientation in the project.

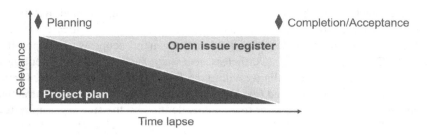

FIGURE 5.9 Dynamics of tasks in the course of the project.

In particular, the process-oriented approach and team structure enabled a way of working that was aligned with the *value streams of* human resource management. This results in the maximum possible encapsulation of activities from one another, which serves the *flow* of processing (results, tasks, contact persons, system areas, etc.).

As usual, as the project progressed, tasks arose that were not previously planned for. Sources for this are unforeseen developments, overlooked necessities, missing decisions, etc., as well as simply the daily events of a complex project. It is crucial to systematically and pragmatically integrate the handling of these tasks into the project processing. For this purpose, the ***open issue list*** served as a central instrument in the PVS project. All new tasks and decisions to be made were administered in this structured register. The open items were recorded, prioritised, scheduled, assigned and documented by the PM team as tasks with regard to their processing status. In the course of the project, the focus gradually shifts away from the plan to the open issue list through the processing of the scheduled tasks (see Figure 5.9) and over time a dynamic ***backlog*** of continuously prioritised elements.

The management of the open items (to-dos, issues) of a project is an indication of the condition of the project as a whole, which I would like to express in the following guiding principle: "Show me the open issue list and I will tell you the state of the project!" In this respect, it has an important significance in the operational management of the project. This instrument was thus seen as another central success-generating element of the PVS project. Important here was the consistent tracking of the elements of the open issue list – especially in the PM team's *jours fixes* – and the ongoing transfer of all logged open items to the register, as shown in Figure 5.10.

In this way, consistent, target-group and deadline-oriented processing was ensured. Ultimately, the achievement of project completion is then operationally equivalent to the absence of high-priority open issues at the end.

However, the described, quite dynamic control elements of the PVS project could not prevent a serious area of waste: The system reporting fixed by project order.

Around 200 reports from the SAP system in the context of personnel administration were contractually fixed. The corresponding requirements were developed by the client in an internal preliminary project and incorporated into the requirements specifications and finally the contract. Accordingly, they were part of the scope of the project and were included in the price. In a SAP system, there are various ways to provide reports: From standard reports, to relatively easy-to-create *queries,* to complex, individually programmable database queries. The 200 required

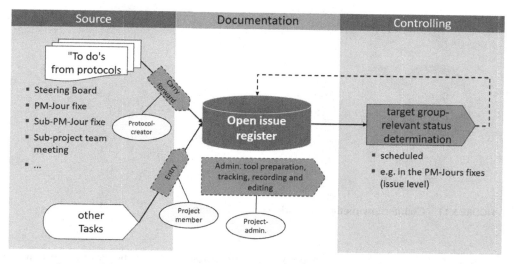

FIGURE 5.10 Success factor open issue management.

reports contained every category, some of which were not really foreseeable in advance with regard to the solution path. The area of reports developed into a point of contention between the project parties – for example, because standard reports originally intended due to requirements in detail now had to be created manually, which entailed a considerable increase in effort. After painstaking work, the contractually required items were finally provided and accepted – only to find out a year after going live that only about 30 of these evaluations were even called up during that period, i.e. actually needed. *Dynamic scope management*, in which the positions are prioritised according to the *MuSCoW principle* would have saved significant resources on both sides (specification, test, re-test etc. respectively conception, development, data preparation etc.). A crucial lesson for Lean PM!

5.1.5 Other elements

APPLIED ELEMENTS OF LEAN PM

- Status Dashboard (Visual Management)
- Task force deployment (Gemba)
- Key figures
- Work package definition as a common core team task (countercurrent procedure, delegation, Hoshin Kanri)

The PVS project was carried out as an overall plan-driven ERP implementation project. This meant that the creation of a ***work breakdown structure*** was a central planning instrument. This essentially includes the identification and structuring of *work packages*, if applicable structured

FIGURE 5.11 Counterflow method of work package planning.

according to *subtasks/projects*. After initial difficulties, the work breakdown structure was realigned in the course of project reengineering in the form of a concerted two-week planning activity in the entire team.

A countercurrent procedure was applied: The central PM team identified the relevant work packages to map to the required scope and to ensure completeness. The planning of the work packages was then carried out in the teams in terms of content, effort estimates, dependencies, etc. This was followed by integrative coordination, in which misunderstandings were uncovered (e.g. on the basis of widely divergent effort estimates), disagreements were resolved and finally working agreements were reached (see Figure 5.11).

With this overarching and ***participatory approach to*** planning, it was finally possible to achieve the desired stability in the processes of implementation, including resilience to change.

A much-used principle of Lean Management is *Visual Management,* i.e. the visualisation of contexts, statuses, etc. This was used intensively in the PVS project, and is illustrated by the ***status dashboard of*** the roll-out (see Figure 5.12).

The status achieved after each time box was successively entered in the fields of this matrix using traffic light symbols: Green = successfully completed; yellow = still in progress; red = conflict; white = not processed. With the vertical view of the dashboard, the planning of the phases could now be updated and tracked team by team. With the horizontal view, it was possible to analyse across teams, i.e. at the level of the central PM, whether, for example, further work on items that had not yet been completed could be suspended by analogy between the teams. The price for this was the (small) risk of a technical misdevelopment, the gain was the planning, effort and deadline reliability.

In the project, ***task forces*** were also formed ad hoc from the project management and technical experts due to the red notifications, which were able to quickly work out solutions against obstacles outside the usual project organisation that the actual team alone would not have achieved.

The dashboard was complemented by the tracking and visualisation of corresponding ***progress indicators*** (see Figure 5.13).

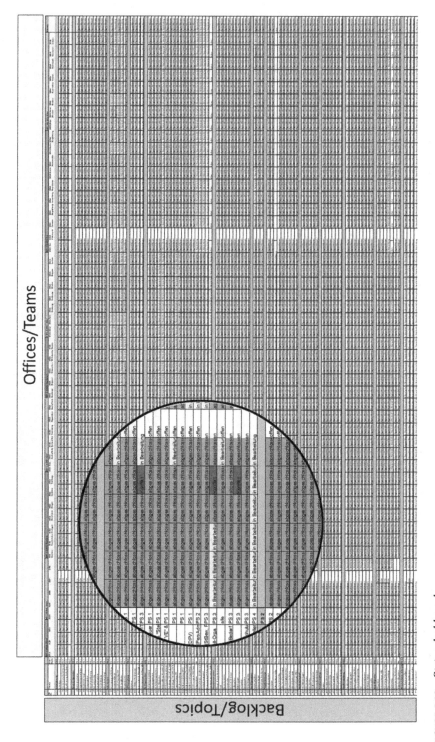

FIGURE 5.12   Status dashboard.

- vertical view (processes) -

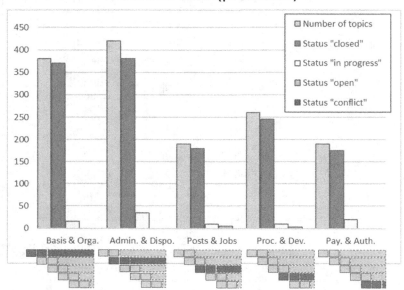

- horizontal view (organisational units) -

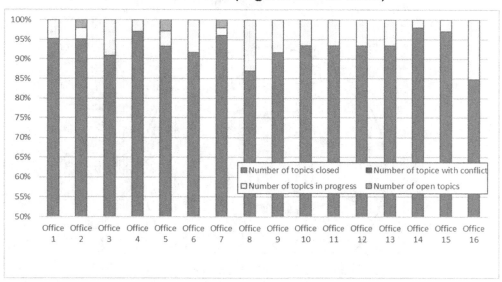

FIGURE 5.13 Key figure-based, visualised status tracking.

IMPORTANT PRACTICES OF THE PROJECT – SUMMARY SELECTION

- Tandems of clients and consultants (*Pair Working*)
- Open issue management through ...
  - ongoing (re-)prioritisation of open items (to-dos, issues etc.)
  - targeted addressing and management of open items
- Time Boxing/Pulsing and Increments ...
  - increase planning reliability with regard to staff availability
  - allow for a high "flow rate"
  - avoid expenditure on things that are not primarily important
- Visualisation through dashboards for targeted intervention
- Delegation of responsibility ...
  - from the business line into the project
  - from the steering committee to commercial project management
- Frequent, regular Lessons Learned etc.

5.2 Case study 2 – ZLP (Central Police Control Centre)

5.2.1 The project at a glance

In another major project, I worked with my team as project manager in a state-wide organisational/ IT/infrastructure project. The "ZLP project", i.e. the establishment of the central control centre of the police of a federal state, was a core outcome of the police structural reform of this federal state. The technical implementation required, above all, the cooperation and coordination of a wide variety of authorities, institutions and service providers. The sub-projects included the conception and implementation of a control centre concentrator, a central switching and communication system, the control centre equipment and the operations control system (ELS) as the core of the application. The second step was the countrywide roll-out, i.e. the successive connection of the decentralised police inspectorates to the system. Project management was carried out in a third-party approach using the phase-based PM method PRINCE2.[5] The project duration of the implementation project up to the productive implementation of the first release was a good 18 months, with a project volume of more than 10 million euros, of which approx. 8% was for PM/project management. A total of 70 workstations for the emergency call dispatchers (360,000 emergency calls per year) as well as 500 concurrent users were to be set up. The specifications comprised 1,200 requirement items. The system created consists of a test/training system and a productive system:

- 230 physical servers and 6 database systems
- 102 virtual servers

---

5 The implementation of the project and the commissioning by the State Minister of the Interior were accompanied by numerous state media such as MDR. The German trade journal Projektmanagement aktuell also published a detailed project report in its February 2014 issue (see Steeger 2014).

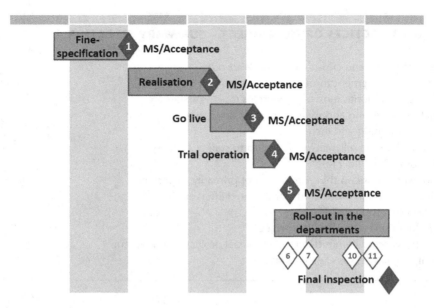

FIGURE 5.14   Overview of project phases.

- 112 active network components
- 106 clients
- Operation in 3 technical centres and 2 digital radio switching centres
- 4 video walls with 36 large-screen monitors

CHALLENGES OF THE PROJECT

- Technological complexity through a mix of infrastructure, hardware, software and organisational components
- Technical criticality with regard to availability and fail-safety
- Dependence on the timely development of external trades
- Extensive requirements catalogue
- Politically motivated time pressure (upcoming state elections)
- Public interest and perception

The project followed a classic phase model, as shown in Figure 5.14.

Many PM elements of the ZLP project were similar to those of the PVS project, so that the following focuses on the differentiating aspects.

5.2.2 Scope management

APPLIED ELEMENTS OF LEAN PM

- MuSCo(W)-Principle
- Systematic change request management
- Release-based engineering

A basic principle of PRINCE2 is to control the project by setting tolerances,[6] which creates room for manoeuvre for the operative PM. Ultimately, these should lead to the next higher instance only having to be called in if these time, monetary or content-related tolerances are exceeded. In the ZLP project, this was consistently implemented, which already began with the design of the requirements. Here, according to an adapted *MuSCo(W) principle,* the extensive requirements were classified according to mandatory and optional requirements, which in sum defined the *minimum degree of fulfilment.* In order to obtain the *breathing scope* which the PM can subsequently use for operational design, it was important that not only the mandatory but also the optional requirements were priced into the agreed financing framework. In the end, this led to a priority-oriented release planning and a change request management system that was capable of acting.

In this *release-based engineering*, which was not developed *a priori* but in detail during the course of the project, the solution could be configured for the first productive start on the basis of the classified requirements in such a way that the technical requirements necessary for operational use were fulfilled. This included the organisational implementations, the emergency call answering, the disposition of the forces, etc. Even if not in complete methodological consistency, the idea of an *MVP approach* was implemented (Minimum Viable Product/Project). The completion of the solution, e.g. simplified operation, extended technical integration, etc., was then carried out in the subsequent release. Last but not least, the findings of the first operating phase were also incorporated.

A second major impact was the handling of *Change Requests* (CR) with the help of the breathing scope. In total, more than 60 change requests were identified and processed in the course of the project until the first go-live. About half of these were accepted and implemented, the rest were rejected. The change requests resulted exclusively from the findings of the technical solution, that were gained during the course of the project. A large number of new requirements for the system to be developed were identified, which turned out to be useful additions as the project progressed. With the help of the optional items in the requirements catalogue, the new features could be designed to be essentially neutral in terms of their monetary impact by deleting a corresponding number of the original nice-to-have items in order to achieve an overall balance in terms of costs. This resulted in budget compliance while maximising utility.

---

6 cf. e.g. Hedeman/Seegers 2012, p. 26 ff.

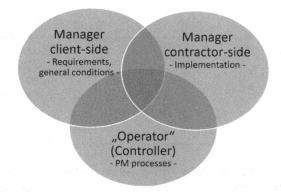

FIGURE 5.15   Schematic representation of the third-party approach with project control.

5.2.3   Project organisation

APPLIED ELEMENTS OF LEAN PM

- Tandem formation (Pair Working)
- Small steering committee
- Project management as a third-party approach (role model according to Unified Project Management Framework)

With regard to the project organisation, many elements that were also used in the PVS project could be transferred. These included, in particular, the *formation of tandems* and the *Small Steering Committee* (see Section 5.1.2). The positive effects were the same, so they will not be discussed further here.

A special feature of the project was the *third-party approach*, in which a *project "operator"* (controller) was entrusted with the task of the operative, especially process-related PM. Such an approach is hardly ever found in the world of IT projects, but is commonplace in the field of construction projects, for example.[7] "Why not learn from it?" was the client's thought when he realised that there was naturally no competence in managing a large IT project in the ranks of the police. In order to achieve a certain independence from the implementing general contractor, the neutral project (process) management, which nevertheless was also contractually committed to the project result, was called in (see Figure 5.15).

In this way, a division of tasks could be carried out in the PM. The leading project manager of the police was especially committed to the domains of requirements catalogue, personnel, budget and legal, the project manager (products) on the contractor side to the development and implementation of solutions and the project manager (processes) to the design and implementation of the PM processes (controlling, risk management, acceptance, etc.). This means that the PM

7 see AHO 2014.

role distribution proposed in the Unified Project Manage was implemented and those involved were able to contribute to the success of the project with a clear distribution of tasks according to their competence and capacity. Last but not least, as project managers (processes) we often acted as moderators and mediators in settling the (unavoidable) conflict situations.

5.2.4 Project planning and control

APPLIED ELEMENTS OF LEAN PM

- Adaptive project controlling (adapted Earned Value Management system, burndown charts)
- Scrum practices such as daily stand-up etc. in the test phase
- "Time before Scope!"

One aspect of planning in practice has already been mentioned – rolling release planning. In general, the project was carried out classically with a multi-stage milestone plan and phase-related, rolling wave detailed planning. This means that, as is usual in a PRINCE2 project, the phase transitions were used specifically to complete the old phase and to prepare the new phase in detail in terms of planning. Due to the strategic-political external pressure – after all, the project was a political issue in the Ministry of the Interior due to its state-wide significance – the overriding motto in the project was to keep to the planned date of going live with only a small tolerance of two to three weeks. Technically and organisationally, an incremental commissioning of parts of the solution was not possible, at least in the first release, so that created a respective pressure.

In the last phase before preparing for production, testing, practices adapted from the *Scrum agile process model* were used. These included the *war room, daily stand-ups,* and *close involvement of product and (PM) process owners* to enable a short-cycle *inspect & adapt approach.* In this way, it was possible to continuously process the extensive test cases and results in close coordination on an hourly basis, to define and immediately implement measures, to evaluate progress on a daily basis and ultimately to enable the go-live with the relevant functionality.

The *War Room* – a project room set up especially for the phase – was occupied by the developers and the testers (team), the test management (process owner) and the project management (product owner). In this way, the organisational advantages could be used for this six-week phase. The following note on this: Due to the size (scope, duration, distributed team) of the project, such a setting was not possible for the entire duration. Rather, it has proven advantageous to use this setting in the "hot phase" of the project, in order to follow the motto *"Deadline (adherence) comes before scope (completeness)!"*

In this phase, a daily *burndown chart* was also created based on the numerous prioritised test cases (see Figure 5.16).

Since the criticality of the test cases was rated more important than the effort required to process them (incoming emergency calls to the police must be able to be dispatched in any case), an effort-related measure such as *story points* was not used here, but rather the pure number of items plus a consideration of defect categories.

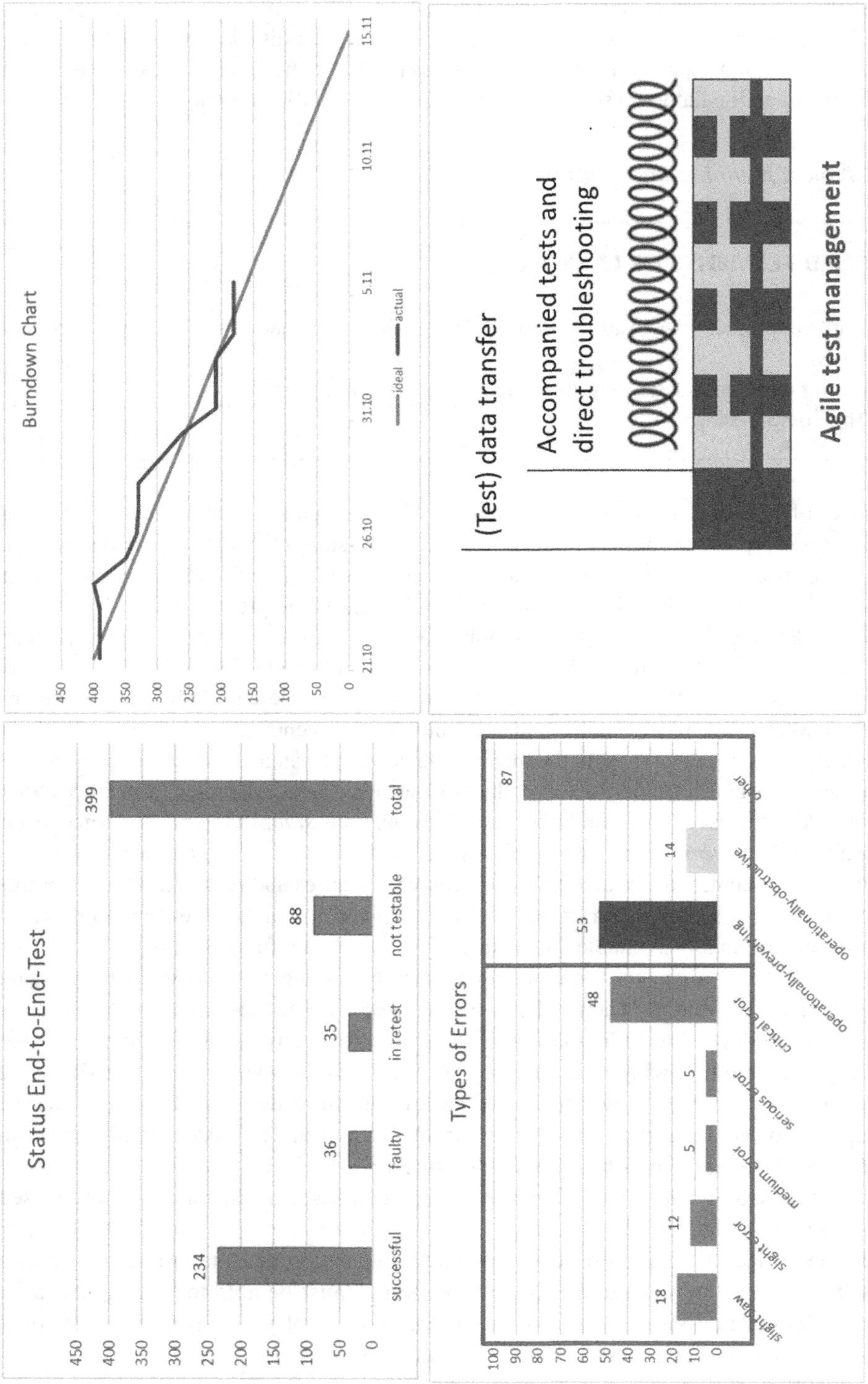

Burndown Chart

Status End-to-End-Test

Types of Errors

(Test) data transfer

Accompanied tests and direct troubleshooting

Agile test management

FIGURE 5.16 Controlling the test progress (section of the end-to-end integration test).

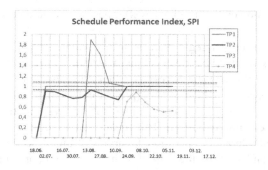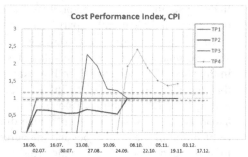

FIGURE 5.17   Adjusted Earned Value Analysis.

During the rest of the project, ***progress tracking and forecasting*** was also carried out using an adapted standard method. *Earned Value Analysis* (EVA) is a professional instrument of project controlling, but it requires certain preconditions if it is to be used beneficially.[8] In a project such as ZLP, the core costs in the project consist in particular of the cost types personnel and material costs. A large number of project staff work with a significant amount of high-priced software and hardware, which is charged over the course of the project. In the cost controlling of the project, the cost types must therefore be considered in a differentiated manner. Only the personnel costs are suitable for use in the Earned Value Analysis. The key figures of the Earned Value Analysis, such as *Actual Cost, Planned Value,* etc., may only refer to the personnel costs and must be extracted from the cost controlling. On the other hand, the degree of completion in such a project is significantly dependent on the availability of the infrastructure. Consequently, an appropriately adapted ***Earned Value Analysis*** was used in the ZLP project (Figure 5.17).

The Earned Value Analysis was part of the standard reporting, which was carried out every 14 days (*jour fixe* and status report). It is also an example of a method adapted to the specific project context.

5.2.5   Mission orientation

APPLIED ELEMENTS OF LEAN PM

- Exception reports
- Business case orientation incl. Benefit Revision Plan and key figures

As already mentioned, in a project carried out according to PRINCE2 the principle of *management by exception* should be practised. In the ZLP project, this principle was implemented, even if not in its pure form. In accordance with the control requirements and the information needs of the respective management level, the core of the reporting system was cyclical reporting. Operationally in a 14-day, strategically in a 3-month, respectively milestone-oriented rhythm.

8 see e.g. Marx/Klotz 2020.

In doing so, the self-imposed premise was followed that (higher-level) milestones should never be more than three months apart.

In the event of significant deviations from the strategic plan or, in particular, threats, the strategic management level (presidents) was involved by means of an *exception report* written in the style of a decision paper. For example, this became necessary when it became apparent that the nationwide geo-data information system, which is the (project-external) technical prerequisite for a central disposition of the emergency forces, might not be ready in time according to the milestone plan, which did eventually happen. This meant that a redesign and re-planning of the project beyond the agreed tolerances/operational margins of the project had to be decided upon, amending the project mandate. With the help of such exception reports, the strategic level could be involved in a measured and targeted way.

The basis for any decisions on changes, whether operational or strategic, was the approved *business case* of the project. This is not so much a monetary cost-benefit analysis, but rather a strategic benefit analysis that justifies the project mandate (according to PRINCE2). Accordingly, it was also possible and sensible to prepare such a Business Case in the regulatory context of the project, the state-wide police structural reform. Based on this, the guiding question "Does the proposed change serve the business case? Does it need to be changed?" could basically guide all decisions on the requirements change management. This is where the aspect of the benefit orientation of the project client – in the case of the ZLP project, the Ministry of the Interior and Police Headquarters – comes into its own.

The other side of benefit planning is benefit collection. In other words, answering the question of whether the project was actually able to realise the intended benefits in the subsequent operation. For this purpose, a *Benefit Revision Plan* was drawn up and adopted in the ZLP project. In addition to qualitative benefit expectations, e.g. the targeted deployment of police forces, quantitative elements also came into play here. For example, the organisation expected the use of the central control centre to reduce the number of decentralised police dispatchers, who could thus be more active in operational police operations and thus increase the capacities available for this purpose – at the same personnel costs. Corresponding *key figures* were identified in the Benefit Revision Plan and evaluated in terms of their measurability. It should be noted that the expected benefit effect of a project can usually only be observed with a significant time delay after going live (i.e. project completion). The reasons for this are settling times and project-related temporary additional expenditure. Here, half a year to a whole year was expected. Therefore, the responsibility for this time-delayed analysis had to be specifically listed in the Benefit Revision Plan and anchored in the operational organisation.

5.2.6   Other selected PM disciplines

APPLIED ELEMENTS OF LEAN PM

- Systematic risk management (register, process, radar, manager)
- Active stakeholders policy
- Lessons Learned after each phase
- Big Picture for the visual representation of the project

ID	Riskname	Description/Indication	Impact	Probability [%]	Impact [%]	Risk Value
1	System acceptance	Insufficient involvement of the user LEZ. Insufficient or late involvement of LPI/Plen.	Process/system not accepted	low (10 - 30%)	remarkable (10 - 30%)	17
2	Political influences Ministry of the Interior	Uncontrolled influence by Ministry of Interior on deadlines/scope of delivery.	Time delay	quite probable (30 - 70%)	remarkable (10 - 30%)	56
3	Building measure LPD	Delays in construction activities result in inability to deploy LEZ on schedule	Time delay	quite likely (30 - 70%)	catastrophic (70 - 100%)	319
4	Building measure Landes RZ	Delays and/or frozen zone jeopardizes scheduled setup of test and/or production environment	Time delay	quite likely (30 - 70%)	catastrophic (70 - 100%)	319
5	Geodata information system	Geospatial information system cannot go live by launch date. Tender has only just been issued. In addition, it is not a standard system but must first be developed.	Quality loss		0%)	56
6	Personnel availability LPD/LKA	Personnel in the participating organizations of the Thuringia police are not available or only to a limited extent. Expertise is lacking or must first be built up	Quality loss		0%)	96

FIGURE 5.18 Risk register (excerpt).

A supreme discipline of PM is risk management. Projects fail because circumstances occur that were not planned for. Especially in large projects with many different organisational, technical and political elements, it is indispensable to handle at least the known uncertainties with the help of systematic risk management. However, a well-established risk management can also provide stability with regard to the unknown uncertainties, such as the unforeseeable outbreak of an epidemic that leads to serious effects on operational activities. This was also the case in the ZLP project.

The operational core of the project's *risk management* was the maintenance of a *risk register*, i.e. a structured list for recording, assessing, allocating and tracking of identified risks (see Figure 5.18). This was to be an integral part of the cyclical project reporting, which was characterised by fortnightly status reports from the sub-projects and, in particular, a *jour fixe* for the PM team with the sub-project managers. It quickly became apparent that the fixed agenda item of *risk updates* every 14 days was perceived as bureaucratism, as the change dynamics of the risks were not correspondingly high. Consequently, the rhythm of the risk update was changed to every four weeks – with good acceptance by the participants and, not least, adequate engagement with the issues ... including avoiding waste.

The success factor for operational risk management was the definition and consistent pursuit of *measures* for the individual risks. These were defined according to risk assessment and monitored on a regular basis.

A tactical element of risk management was the appointment of *risk managers* from the ranks of the (sub-)project managers. Risk categories were formed and a manager was assigned to each of them. The categories included time/schedule, quality/scope, infrastructure, personnel, budget and general conditions. The risk managers were in charge (implementation responsibility) of the processing as well as the risk identification and monitoring of the risks of the assigned category. The project manager retained overall responsibility for risk management. The risk managers were appointed in such a way that the respective person had an organisational or professional connection to the corresponding risk category, so that maximum competence and also access to corresponding information (*risk radar function*) was achieved. In this way, not only the identified risks, i.e. the known uncertainties ("known unknowns"), but also the unexpected ("unknown unknowns") could be handled in a stable manner and the project helped to achieve resilience.

The risk management of a project should always be linked to systematic stakeholder management. This is also the case with the ZLP project (risk category framework conditions). Influential stakeholders can become a risk for a project and should be actively managed. In the ZLP project, an *active stakeholder policy* was therefore pursued by actively communicating with police departments, the state parliament, politicians and the regional press. As an internal and security policy measure of the state, the ZLP project attracted increased attention from the institutions mentioned, whose need for information and participation was professionally met with the help of the press office of the state police authority.

*A **Big Picture** of* the project was also developed as a methodical element of communication. With such a visualisation, the scope, organisation, procedure, boundary conditions, etc. can be clearly summarised, explained and discussed. The Big Picture of a project thus contributes very well to the communication of the project internally and externally.

Learning from experience – this maxim is one of the success factors of modern PM. In the phase-oriented ZLP project, retrospectives in the form of ***Lessons Learned*** workshops were therefore conducted with project participants at each phase transition (see Figure 5.19).

In the workshops, success factors (maintain!) and disruptive factors (abolish!) were systematically identified and measures for dealing with them were derived. Although the team of police staff had to be convinced of this at the beginning, the findings, as expected, proved to be very helpful for the continuous improvement of the project processes in particular. Particularly with regard to the PM processes, it was thus possible to uncover and realise potential for improvement and acceptance at an early stage, even in the fundamentally sequential procedural model of the project. The possibility for team members to be systematically involved in the improvement of work processes in this way should not be underestimated as a measure that promotes acceptance. It is critical for success that the findings are followed by consequences.

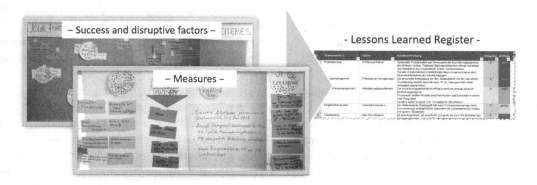

FIGURE 5.19 Documentation and administration of Lessons Learned.

IMPORTANT PRACTICES OF THE PROJECT – SUMMARY SELECTION

- Project management as a third-party approach
- Tandems of clients and developers (Pair Working)
- Open-point management through ...
  - ongoing (re-)prioritisation of open items (to-dos, issues, etc.)
  - targeted addressing and management of open items
- Systematic, ongoing risk management
- Adapted Earned Value Analysis and other controlling tools
- Situational, temporary use of selected Scrum practices etc.

6

ANALYSING THE CONDITIONS OF A PROJECT

After reading this chapter, you will know ...

- how to create a project profile for systematic PM system adaption,
- a canon of 18 criteria for classifying a project in terms of a morphological box

Drawing on the experience of the case studies described in Chapter 5, this chapter presents ways of systematically analysing project characteristics in order to select an optimal PM approach or procedure.

> The manager's skill is to know different management ideas and to choose the one that is best suited to the actual challenge in which he finds himself – just as he might choose the right golf club according to the specific position of the ball.
>
> *(freely adapted from John Harvey-Jones)*

In the Chapter 5 two examples were presented in which the project methods applied in concrete projects were adapted more or less situationally and partly reactively. The aim in each case was to use those practices that on the one hand corresponded to current requirements (such as delivering results in the short term) and on the other hand naturally corresponded to generally known good practices (such as tandem formation), i.e. practices were used ad hoc or a priori. Based on this insight, the question arose as to how this approach can be systematised so that projects in general can benefit from these considerations.

DOI: 10.4324/9781003435402-7

6.1 Classification of projects

In addition to the inherent challenge of projects being unique and novel undertakings, which are therefore genuinely fraught with uncertainties and risks, the empirical analysis shows that the contextual circumstances of a concrete project are often not taken into account in the sense of the specific design of the project's PM system.[1] In other words, the management requirements of projects fluctuate on a case-by-case basis and should be reflected in an adaptive design.[2]

Basic and fundamental systemic project characteristics such as complexity, uniqueness and time limits are not in themselves an indication of the type of project and the specific characteristics it entails. For such a classification, further aspects must be considered.[3] Due to the multifaceted characteristics of projects, different project types can be identified, which also place special demands on the PM according to their characteristics.[4] A typification of the project in the early planning or initiation phase can be decisive for a successful course of the project, as this provides the PM with the first basic information for working out the project design.[5] In particular, this should prevent projects from requiring higher financial and time resources than initially planned – a scenario that often occurs in reality. Practical examples, such as the construction of Denver International Airport (1989), underpin the vital importance of analyses with regard to project typing and a comprehensive level of knowledge before the start of the project.

EXAMPLE DENVER INTERNATIONAL AIRPORT

It was the wish of the city of Denver that the construction of the baggage transport system be divided up, with each airline planning its own system and implementing it independently. It was realised far too late that only one airline complied with this wish. As a result, their system was to be applied to the entire airfield. However, this airline constructed an advanced and automated system, the construction of which was far more complex and time-consuming than envisaged in the overall planning of the project. As a result, a standard large-scale project additionally became an innovation project. This led to delays in the project process and constant changes in requirements, which increased the costs and duration of the project. The misplanning of the baggage handling system alone resulted in additional costs of 118 million US dollars. The opening of the airport was delayed from October 1993 to February 1995. Nevertheless, there were severe functional limitations after the opening, which in retrospect can be attributed to the failed planning.[6]

The classification and typification of projects helps the PM, with the use of various criteria and characteristics, to focus specifically on project initialisation, definition, planning, control

---

1 see Shenhar/Dvir 2007, p. 21 ff.
2 see Gessler 2012, p. 51 and GPM 2017, p. 106.
3 see Kuster et al. 2011, p. 5 ff.
4 see Zell 2017, p. 5.
5 cf. Frick et al. 2019, p. 1006.
6 see Büttgen/Fabricius n.d., p. 1 and Drexl et al. 2002.

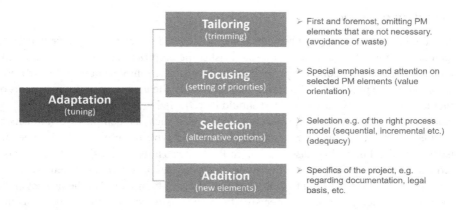

FIGURE 6.1 Clarification of terms for the adaptation of a PM system.

and project completion. Since different PM tasks have different degrees of importance within a project, they must be differently weighted and processed accordingly. By creating and analysing the project profile at an early stage and critically examining the resulting measures, risks can be identified and, at best, eliminated or controlled in order to achieve the desired project result.[7]

Some of the well-known PM standards are so extensive – for example the current version of GPM's PM Compendium comprises almost 2,000 pages – that they are often perceived as bureaucratism, i.e. waste, if applied rigidly. On the other hand, they are also designed as a body of knowledge and thus generically, which does not even claim to be directly operationally applicable, but immanently requires the step of operationalisation. Many of the well-known PM frameworks more or less explicitly include the demand for a specific design of the PM system. Examples of this are PRINCE2 and the V-Modell XT. PRINCE2 calls for "adaptation to the project environment" within its seven *basic principles*. As influencing factors, PRINCE2 mentions the industry, existing company standards, the PM maturity of the organisation as well as factors such as size, complexity, importance, capacity and risks.[8] The latter influencing factors are also included in the V-Modell XT. V-Modell XT refers to the adaptation of the standard to project-specific needs as tailoring. In this process, the products, activities and general elements that are not needed in the project based on its typification are omitted. In principle, however, the addition of supplementally required elements is also permitted. In addition to the factors mentioned above, the subject of the project (e.g. system development vs. system adaptation) is used for typification.[9]

In the following, the model shown in Figure 6.1 will be used to illustrate the ***adaptation of a PM system***. PM system refers here to the management of a concrete project, as opposed to the PM system of a whole organisation.

---

7 see e.g. PMI 2017, p. 395 ff.
8 see AXELOS 2017, p. 64 ff.
9 see Höhn et al. 2008.

In the course of adaptation, the PM disciplines must be (1) neglected, (2) strengthened, (3) given an alternative design or (4) supplemented compared to usual approaches. Examples of this are:

1. The construction of a new warehouse is an investment project. Change management can tend to be neglected (tailoring in the narrower sense).
2. The merger of two companies is an organisational project. Special emphasis must be placed on change management (focus).
3. The innovative design of the customer journey with the help of a customer portal is a marketing innovation project. The approach should be incremental (agile) (selection).
4. The roll-out of an accounting IT system to China requires compliance with special regulations not provided for in any standard. These must be added to the procedure model accordingly (supplement).

Depending on the project context, there are different elements of PM that are critical to success. If these are not identified and taken into account, the chance of success is reduced. If, on the other hand, capacity and energy are wasted on elements that may not be so critical to success, then the elements that actually make sense will inevitably suffer.

The result is a hybrid, project-specific PM system. *Hybrid* refers to the use of elements of different standards or best practices (see definition in Section 1.2). This includes, in particular, the mixing of classical, plan-driven and agile elements.

DIN 69 901 defines a PM system as a system of guidelines, organisational structures, processes and methods for planning, monitoring and controlling projects.[10] The reference to the organisation as a whole or to the individual project is left open. The Project Management Institute (PMI) clearly assigns the term to the level of the individual project: According to this, a PM system is the "combination of processes, tools, methods, methodologies, resources and procedures for the management of a project",[11] which is documented in the project-specific PM plan (comparable with a project manual). In an earlier version, the PMI adds that the elements must be "consolidated and combined into a functional, unified whole".[12] In the context of Lean PM, we adopt the view of a single project. In contrast, a PM system that presents itself as a higher-level, organisation-wide management system should be called an *enterprise* or *corporate PM system*. The CPM system is thus adapted to the requirements of the individual project.

DEFINITION PM SYSTEMS

- A PM system is a system of structures, processes, practices and guidelines for managing a project.
- A corporate PM system (CPM system) is an organisation-wide management system for initiating, executing and following up on the elements of an organisation's project landscape. It has a normative effect on the PM systems of individual projects.

---

10 DIN 2009b, p. 14.
11 PMI 2017, see 715.
12 PMI 2004, p. 372.

The CPM system is therefore the regulatory "umbrella" within which the PM systems of individual projects can be designed.

6.2 Project characterisation criteria

6.2.1 General alignment of the PM system

In order to carry out this goal-oriented adaptation of the PM system, the question arises as to which criteria can and should be used for this purpose, i.e. which characteristics of a project are relevant in this respect.

On the way to project-specific characterisation in a company, an organisation-wide classification should first be carried out. In this classification, the company's projects are divided into categories according to a defined, company-specific taxonomy. These categories consider, for example, the size of the project (monetary as well as organisational) or the scope in the organisation (division-wide or company-wide). This categorisation is thus directly part of the design of the CPM system and is accordingly discussed in Chapter 8.

At this point, on the other hand, we are primarily concerned with the individual classification of the project in order to configure the supposedly optimal PM system for it. In the literature, several approaches can be found for classifying projects with regard to the orientation of the PM system. Pioneers are Shenhar and Dvir, who presented a four-dimensional, novel system for this task with their *Project Diamond* ("Reinventing Project Management").[13] In this system, the dimensions of influence *complexity, technology, novelty* and *speed* are proposed, which were determined empirically. The following characteristics of these parameters are given (each with ascending demands on the PM):

Complexity with regard to the project scope:

- Assembly – creation of a single function or a subsystem of a component, if applicable by merging already existing parts
- System – the development of a complex system
- Array – complex system, which in turn consists of multiple, interacting systems

Technology with a view to the associated uncertainty:

- Low-tech – known, mastered technology
- Medium-tech – mainly known technology, with some new elements
- High-tech – predominantly new, but already existing technology
- Super-high-tech – new technology not yet available at the time the project starts

Novelty in the sense of innovation for customers:

- Derivative – extensions and improvements to existing services
- Platform – new series of a basically already existing product
- Breakthrough – new products with previously unknown market

---

13 Shenhar/Dvir 2007.

Speed (Pace), i.e. the urgency of delivering the result:

- Regular – uncritical
- Fast/competitive – the sooner available, the better
- Time-critical – provided with an (external) deadline
- "Flash" – a quickly brought about result to resolve a crisis situation is imperative

Neither standard procedures nor standard categories exist for project typing. The existing concepts have their raison d'être; from them, insights for PM can be gained from the characteristics of the projects. However, they do not appear to be complete. For example, employees often perceive the project outcome of organisational projects (e.g. changed work processes) as threatening and resist this type of project.[14] Here, a stronger focus should be placed on change management in order to take away employees' fear and win them over to the project. Project types also differ in their management. For example, in organisational projects and R&D projects, in contrast to infrastructure projects, the progress of the project can only be inadequately evaluated.[15] The monetary evaluation of external and internal projects is similarly conflicting. Each of these project types offers insights for successful PM.

In the following, a model for project typing is created,[16] so that users can create a project profile taking into account various criteria. Characteristics that are relevant for project typing or for profile creation and analysis have been identified and are presented including their characteristics.

Complexity in terms of subject matter and content: Degree of interconnectedness of different systems. For example, the complexity of technical systems can merge with the complexity of management to form a very complex structure with high momentum. The literature shows that complexity cannot be measured exactly but can be assessed by experts and specialists.

Complexity in terms of social matter and communication: Personnel size of the team involved in the project. The larger the project team and the number of stakeholders, the more coordination difficulties arise and the greater the administrative and coordination effort. The social-communicative complexity increases.

Commitment: Expected attitude towards the project (motivation) – both among the clients and the staff. With improved commitment, fewer conflicts can be expected and vice versa. Depending on the degree of commitment, resistance and fears on the part of employees must be overcome by managers or the respective project leader.

Experience: Availability of the necessary knowledge in the company to deliver the project objective. With increasing experience, ideally standards written by experts are available and can be applied. With decreasing experience, personnel measures are associated, e.g. training of employees needed for the project.

Interdisciplinarity: Number of different specialist areas (disciplines or organisational units) involved in the project process. As the degree of interdisciplinarity increases, more interfaces

---

14 see Ottmann et al. 2008, p. 36.
15 see Ottmann et al. 2008, p. 37.
16 cf. here and in the following Gessler 2012; Felkai/Beiderwieden 2011; Kuster et al. 2011; Shenhar/Dvir 2007.

and thus more potential communication conflicts arise. These need to be analysed in more detail.

Geographical distribution of the project team: Describes how many different places the individual members of the project team are located in. The different characteristics provide information about how far-reaching the distribution is. The higher the level, the more decentralised the distribution of the project team, and the greater the spatial distances and, if applicable, temporal shifts.

Personnel effort: Influence of the project on the day-to-day business and the resources required for it. With increasing effort, day-to-day business is negatively affected, so that, e.g. analyses and measures have to be initiated.

Time expenditure: The duration of the project in the narrower sense and, in the broader sense, its influence on day-to-day business and the resources required for it. With increasing effort, day-to-day business is negatively affected, so that, e.g. analyses and measures have to be initiated.

Financial cost: Investment level of the project and the possible financial burden on the company. With increasing financial expenditure, the company can be negatively affected by the project. As the investment level increases, target/actual comparisons must be carried out more frequently by the project management in order to be able to initiate measures in time.

Client: Position of the client with respect to the project. External clients are usually customers. Depending on their characteristics, the stakeholders in the project are to be considered differently, so that derived precautions and measures correspond to the respective client and other stakeholders.

Intangible output: Intangible share of the total output. As this criterion becomes more pronounced, the influence on the project outcome becomes greater. In this context, documentation or communication can play a more important role in planning as the intangible share for the PM increases.

Material output: This criterion describes the material share of the total service production. If the material share of project output increases, the logistical effort with regard to materials and the goods to be produced increases, for example. Thus, an increased focus can be placed on supplier management in PM.

Quality requirements: Quality describes the degree to which a defined set of characteristics of the object of consideration fulfils the requirements placed on it.[17] In this context, the quality requirements indicate the criticality associated with defect tolerance in relation to the project result. For example, a project in the field of aircraft construction has zero defect tolerance. Controls and measures of the project must be aligned with this. Therefore, with increasing criticality, additional time and monetary resources must be expected in the PM. They must be planned for in order to meet customer requirements. In addition, the risk associated with the project increases.

Urgency: Urgency describes the time pressure in the project. It reflects the deadline within which the service or solution must be provided. If there is a high degree of urgency, further

17 see DIN 2015, p. 39.

measures can be planned in the PM in order to achieve the result more quickly. This is usually associated with high financial and/or personnel effort.

Strategic importance: This criterion describes the overriding importance of the project, with regard to individual divisions or the entire company. With increasing strategic importance, the influence of the project results on the company's development increases.

Degree of innovation: The degree of innovation describes the circumstances, instruments and activities to be expected or occurring in the project, the nature of which are novel and unfamiliar. According to the novel elements, measures should be taken in the PM to reduce risks arising from this novelty. If, for example, the use of a new type of instrument is planned in a project, appropriate measures for training the staff should be planned.

Plannability: Plannability describes the possibility of planning or subdividing the course of the project into concretely defined sub-steps. The higher the level, the more precise project planning can be realised. The more precisely a project can be planned, the better, for example, target-performance comparisons can be made, and appropriate measures can be initiated.

Technology level: This criterion describes uncertainties and expected difficulties in connection with the technologies to be used. In some cases, the solution path may also be undefined. In a narrower sense, this criterion is characterised by the technology level at which the service creation process is taking place. The higher the technology level, the higher the associated uncertainties or difficulties.

The criteria described are shown in Figure 6.2 in an overview.

The project profile resulting from the morphological box serves as a basis for the design of the PM and can be used in relation to phases. Critical areas can be identified and measures derived from them. Compared to a rigidly defined project process, these measures can be shifted to earlier or later project phases. When assigning the respective value characteristics, it should be noted that these are to be considered relatively. The criterion of financial expenditure can be used as an example. The financial effort of a project can mean a high burden for company A, whereas company B is only slightly affected by it. The companies will therefore evaluate the characteristics differently.

Subsequently, the respective project profiles are analysed – for example via a network diagram – and recommendations for action are derived from them, as shown in Figure 6.3 in principle.

In Figure 6.4 the consequences for the design of the PM system are proposed. These, as well as the criteria themselves, cannot claim to be complete, since due to the definitional uniqueness of projects, an individual assessment should also be made in each case.

6.2.2 Case study: Creating an audio tour

A technology museum is opening a new exhibition on the subject of *aircraft in the First World War*. In addition to the pure display of the exhibits, an audio guide is to be created to guide visitors through the exhibition and provide various additional information. This auditory support has to be worked out, recorded and played on the audio devices. Furthermore, additional loudspeaker units are to be distributed within the museum to create the most impressive experience possible. The project profile created by the Museum of Technology is shown in Figure 6.5.

Criterion	Minimal occurence	Medium occurence	Maximum occurence
Complexity (technical-contentual)	Clear, low number/interconnectedness of elements with little dynamism	Medium, manageable	High, not manageable, strong networking of elements with a lot of dynamics
Complexity (social-communicative)	Few participants, low administration/coordination effort	Medium number of participants, increased administration/coordination and support effort	Many participants, high administration/coordination effort, great dynamism
Geographical distribution of the team	Entire project team in close proximity, local	Distributed at several locations, regional	Distributed at many locations, supraregional, possibly international/global
Commitment	Good, few conflicts expected	Varies, requires management attention	Critical, crises and resistance to be expected
Experience	Existing (routine project), standards in place	Partly available among key persons, few standards	Not available, must be built up, no standards
Interdisciplinarity	One department	Few departments	Many departments
Effort, personnel	Low, does not influence day-to-day business	Medium, can be implemented with existing resources, influences day-to-day business	High, requires additional resources, day-to-day business is interrupted
Effort, temporal	Low, does not influence day-to-day business	Medium, can be implemented with existing resources, influences day-to-day business	High, requires additional resources, daily business is interrupted
Effort, financial	Investment amount Low, insignificant	Investment medium, requires certain precautions	High, requires additional resources, day-to-day business is interrupted
Client	Internal client	Client in affiliated company, e.g. within a group of companies	External client
Service provision (tangible)	No/low share, insignificant	Medium, requires arrangements	High proportion, project outcome highly dependent
Service provision (intangible)	No/low share, insignificant	Medium share, requires precautions	High proportion, project outcome highly dependent
Quality requirements	Low criticality, little additional resource consumption, low risk	Increased criticality, moderate additional resource consumption, medium risk	Maximum criticality, pronounced additional consumption of resources, high risk

FIGURE 6.2 Morphological box for project characteristics.

Criterion	Minimal occurence	Medium occurence	Maximum occurence
Urgency	Low, meeting deadlines not critical	Medium, deadline fixed, sufficient time available	High, deadline fixed, time critical
Strategic importance	Low, strategically not significant	Medium, strategically important for individual business units	High, strategically important for the entire company
Degree of innovation	Low, reference-based, routines and standards in place, low risk	Medium, building up, novel elements, medium risk	High, radical, no standards, high risk
Plannability	Good, clearly plannable	Moderate, difficult to plan	Poor, almost impossible to plan
Technology level	Low-tech, no uncertainties or difficulties to be expected	Medium-Tech, moderate uncertainties or difficulties to be expected	High-tech, strong uncertainties or difficulties to be expected

FIGURE 6.2   (Continued)

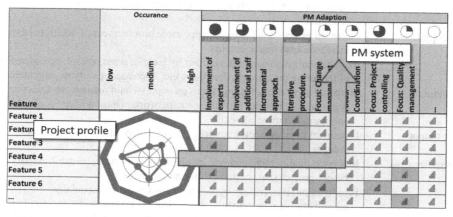

FIGURE 6.3   Derivation of recommendations for action.

Various aspects and special features of the project can be seen from the profile. The company can draw on a wealth of experience in this project due to previous exhibitions. A routine approach is therefore possible. Furthermore, the characteristics of the degree of innovation and the ability to plan are rather low, since similar projects are carried out quite frequently. A standardised procedure is documented in the company. Furthermore, the project is not very complex and the effort for the company is rather low. The intangible part of the service provision in the form of the audio guide is quite pronounced. However, due to the additional loudspeaker units, which are to be installed mainly hidden, there is also a small material project share.

Criterion	Meaningful consequences of action
Complexity, technical and content-related	Iterative and/or incremental approach, high-level planning, sensible reduction of the scope, subdivision (sub-projects), integration and architecture management
Complexity, social-communicative	Distinct coordination of & communication to project participants, efficient and organised resource management, more far-reaching planning, subdivision (sub-projects)
Geographical distribution of the team	Distinct coordination of & communication to project participants, strengthened time management, collaboration tool
Commitment	Increased focus on change management, staff training, senior management involvement, alignment with visions
Experience	Iterative approach (learning curve), involvement of experts, establishment of a Lessons Learned process (development of standards), knowledge management
Interdisciplinarity	Strong coordination of & communication to project participants, team-building, stakeholder management, focus on value-adding participation (where possible)
Effort, personnel	Encapsulation of the project (no matrix organisation), no multi-tasking, involvement of line managers, efficient and organised resource management, involvement of additional staff (e.g. freelancers)
Effort, temporal	Realise quick wins, if possible define time-bound project phases with self-sufficient results, involve additional staff (e.g. freelancers), influence deadlines
Effort, financial	Increased project controlling, if applicable procurement of additional capital, increased cash management
Client	Requirements management, contract incl. claim management, consideration of tendering procedures, deadlines, etc., increased quality management
Material output	Increased project controlling, focus on supply chain management, increased resource and logistics planning (e.g. network planning technique), configuration management
Intangible output	Document management, knowledge management, data protection and security
Quality requirements	Increased quality management, involvement of experts, research activity before project start, customer/user participation
Urgency	Prioritisation, if necessary bypass of usual standards, especially administrative requirements, short, clear decision-making and escalation paths, no multi-tasking, integration of additional staff (e.g. freelancers), increased time management.
Strategic importance	Senior management involvement, experience-based staffing of the project team, change management, stakeholder awareness, increased project controlling, more frequent review cycles, increased risk management
Degree of innovation	Iterative and/or incremental approach, prototyping, customer/user participation, minimum viable product, product vision, involvement of experts, joint ventures with other companies, subdivision (sub-projects)
Plannability	Use of agile methods (e.g. Scrum), iterative approach, rolling wave planning
Technology level	Experience-based staffing of the project team, integration management, prototyping, testing, increased risk management, more intensive cooperation with customers, involvement of experts, joint ventures with other companies

FIGURE 6.4 Overview of the consequences if the criterion is met at the maximum level.

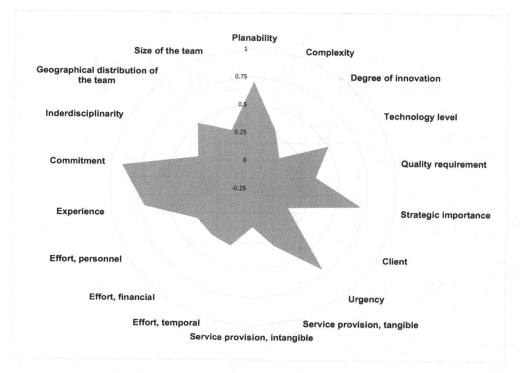

Planability

Size of the team

Complexity

Geographical distribution of
the team

Degree of innovation

Inderdisciplinarity

Technology level

Commitment

Quality requirement

Experience

Strategic importance

Effort, personnel

Client

Effort, financial

Urgency

Effort, temporal

Service provision, tangible

Service provision, intangible

1
0.75
0.5
0.25
0
-0.25

FIGURE 6.5 Project profile – Creating an audio tour.

A special feature that the museum has to take into account during the implementation is the fixed opening date of the exhibition. Because of this, the project must be planned in such a way that completion on this date can be guaranteed in any case. Nevertheless, the quality of the audio tour must not suffer from possible time-critical processes.

With the help of the generated project profile, recommendations for action are derived, which should then be given greater consideration in the project process. The pie charts in Figure 6.2 above the individual recommendations for action show their overall weighting. Since some recommendations are influenced by different criteria, a stronger weighting can also result from several criteria that tend to be rated low. In this example, it should be examined whether additional experts should be involved. Furthermore, due to the fixed deadline and the urgency at hand, the involvement of additional staff should be considered. In addition, project controlling should be considered particularly important. These recommendations are based, among other things, on the strategic importance of the project for the museum.

Due to the degree of innovation for audio guidance, an iterative approach should be implemented in addition to the involvement of specific experts. The further recommendations for action can be found in Figure 6.6.

With the help of the elaborated system, there is a far-reaching method for project typification for a large number of practitioners. Based on the determined project profile, the focal points of a project can be recognised and taken into account in PM and its sub-disciplines. A basic problem

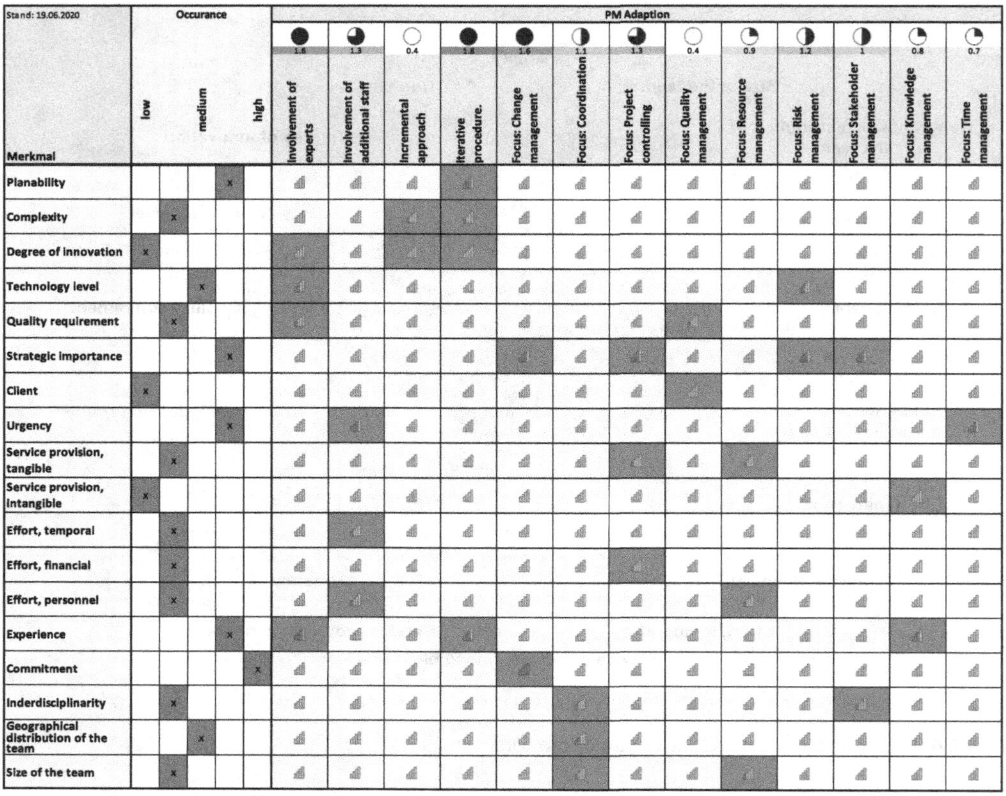

FIGURE 6.6 Overview of recommendations for action – creation of an audio guide.

here is the handling of complexity. An immanent characteristic of complexity is the temporal and structural volatility of the system under consideration (dynamics or emergence), in this case the project. This instability leads to increased planning uncertainty, which puts the focus on the use of agile approaches (see Sections 2.3 and 4.8).

6.3 Overall view of the PM system adaptation

Overall, consideration of the adaptation of PM systems results in the overall system shown in Figure 6.7. The hexagon consisting of project type, category and profile holistically describes the typification of projects for the adaptation of the PM system.

1. First of all, the **project type** spans the basic problem and solution space:
   Each economic, technical or professional domain, e.g. an industry, defines the framework conditions under which a project is to be carried out. These include legal or regulatory constraints or technical requirements that are ultimately induced by the project subject. A construction project runs under different conditions than a project for process optimisation

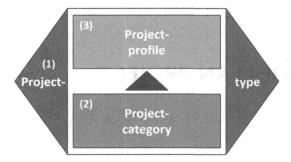

FIGURE 6.7   PM system adaptation strategy.

or a moon mission. The defining feature here is the procedure in the technical development of results.

2. The **project category** then classifies the project in terms of corporate management:
The relative importance of a project ranks it within the company in the context of portfolio management. This expresses the strategic importance and criticality of the project for the company's success. A typical categorisation divides projects into A-/B-/C-projects. This classification already results in essential requirements for the project's PM system, such as the reporting system.

3. The **project profile** concretises the project-specific characteristics based on the context:
Finally, the individual profiling of the project completes the design of its PM system – within the possibilities set by the project type and category. Through the specific project profile, it is possible to identify the technical and methodological characteristics of the project and transfer them into an operational approach, as described in detail in this chapter. Novel, uncertain project objects require different approaches than known, stable, etc. projects.

In general, it should be noted that in the process, each PM discipline should always be implemented at least at a basic level (see Section 1.1.2). This means that the complete omission of entire sub-domains of PM, such as risk management, etc., generally proves to be as inappropriate, as it is not promising.

7

IMPLEMENTATION OF LEAN PROJECT MANAGEMENT

After reading this chapter, you will know ...

- about the influence of organisational and cultural boundary conditions,
- the procedure of implementing Lean PM, focusing on change management aspects, and
- some relevant practices and success factors for the implementation.

In this chapter we look at the implementation of Lean PM in the practice of individual projects and the general establishment of Lean PM in the organisation. The following questions serve as a guideline:

- Is Lean PM equally suitable for all organisations? What are the influencing factors at the level of societal culture, the corporate culture as well as the character traits of the individual persons?
- What aspects need to be considered when introducing Lean PM as a CPM system? What is the most appropriate approach?
- How can Lean PM be applied situationally in projects – even without it being an official CPM approach?

7.1 Organisational classification

7.1.1 Cultural framework

When introducing Lean PM in a company, different objectives have to be considered, which are not mutually exclusive but complementary. On the one hand, there is the fundamental

DOI: 10.4324/9781003435402-8

establishment of Lean PM as a corporate PM system in the organisation. On the other hand, it is about the application of Lean PM in individual projects. Finally, it is about the question of how the ideas and methods of Lean PM should and can be implemented in the organisation.

The establishment of Lean PM as a CPM system goes hand in hand with understanding and, if necessary, creating the overall organisational prerequisites. Here, questions of culture – at the company level, but also beyond that – as well as the individual suitability and competence of the employees are relevant. In any case, the fundamental implementation of Lean PM as a CPM system must be understood as a change process and thus be designed with the methods of change management.

The second level involves the implementation of Lean PM in individual projects. This can be done either top-down, i.e. on the basis of an appropriately designed CPM system, or bottom-up, i.e. on the basis of the operational sense in the concrete project (see Chapters 5 and 6). With a concrete project reference, the operational methods of Lean PM come to the fore compared to (strategic) change management.

There are a number of principles associated with Lean PM, which are linked to the *3Gs – Participation, Pareto principle and Fit* – in the sense of a compact guideline (see Section 3.4). Some of the elements bundled under this can be identified as elements whose implementation requires certain cultural preconditions. These include the participation of the team in planning and decision-making, with the greatest possible delegation of responsibility to the team, as well as the continuous involvement of process and service customers. Furthermore, the black box concept, i.e. the conscious allowance of design gaps, expressed for example in rolling wave planning or a more open-ended design of requirements, requires trust in the abilities of those involved and the possibilities for readjustment in the procedure – including a forgiving culture. Last but not least, the principle of fit requires that the participants have the appropriate competence in technical, methodological and leadership terms. The organisational design and the staff working in the organisation should fit together (see Figure 7.1).[1]

The classification according to Kuwert in Figure 7.1 shows that (only) active designers develop their potential in agile organisational forms, while the type of "well-behaved" implementer is overburdened with a structure based on self-organisation. He prefers an organisation with a pattern of command and control. On the other hand, shadow organisations of people who want to actively shape things will form – bypassing this very pattern, as it is perceived as inhibiting.

Let us approach these aspects one by one. Culturally conditioned issues of leadership, planning, decision-making or task allocation have taken on crucial importance for managers and organisational designers in international and multinational companies in the last few decades, and have been shaped by the globalisation of the world economy. The complexity of merging corporate cultures, in which the characteristics of the individuals involved, shaped by cultural differences, are a factor for success, has led to the development of a number of sociological models that can be used to systematically describe organisational cultures and their individuals. From the multitude of intercultural experts who have developed various dimensions for the

---

1 according to Kuwert 2019.

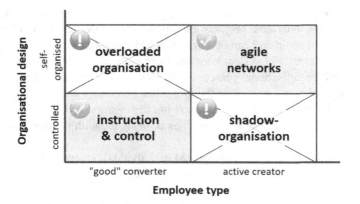

FIGURE 7.1 Fit between organisational design and employee type.

simple description of these aspects, often through extensive empirical analyses, some of the best known are mentioned:

- Hall classifies groups as mono- or polychronic, context-rich or context-poor, and past- or future-oriented.[2]
- Kluckhohn and Strodtbeck recognised the following basic questions and value orientations: Attitude towards time, nature and humanity, relationship to other people, behavioural motive as well as human nature.[3]
- Hofstede's model includes the cultural dimensions of power distance, collectivism vs. individualism, femininity vs. masculinity, uncertainty avoidance and long-term vs. short-term orientation.[4]
- Hamden-Turner and Trompenaars' dimensions are: Universalist vs. particularist, individualist vs. collectivist, specific vs. diffuse, meritocratic vs. ascriptive, and neutral vs. emotional or affective.[5]
- Lewis divides people into three typologies based not on nationality or religion but on behaviour: Linear-active, multi-active and reactive.[6]

The GLOBE (Global Leadership and Organizational Behaviour Effectiveness Program) study, which has been running since 1994, also empirically confirms the assumption of implicit, culturally influenced leadership ideas. According to this, expectations of leaders differ in different cultures and these differences can be traced back to socio-cultural values.[7]

2 see CrossCulture n.d.
3 see Hills 2002.
4 see Hofstede 1997.
5 see Trompenaars/Hamden-Turner 1996.
6 see Lewis 2006.
7 cf. Brodbeck 2016.

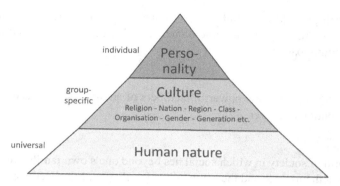

FIGURE 7.2   Levels of mental programming.

A more detailed discussion of these cultural models is beyond the scope of this paper. In particular, an assessment of the models should be left to sociologists and business psychologists. It is indisputable that these models always only depict stereotypes that were usually statistically derived from empirical sociological studies. Hofstede, for example, developed his model on the basis of studies with more than 110,000 IBM employees worldwide (from more than 50 nations and regions).[8] These models are intended to serve as an indication with which different structures in organisations and individual behaviour of people can be better understood. Hofstede describes culture as a *mental programme*, i.e. patterns of thought, feeling and action based on the social environment and life experience.[9] Overall, this results in Figure 7.2[10] where the higher level naturally dominates the lower level in terms of the mental programming of the individual.

In the following, some elements of the cited models will now be used to derive some basic considerations for the application of Lean PM. In particular, the culturally relevant factors identified at the beginning of the 3Gs model. To classify Lean PM in a culturally shaped profile, the three genuine sources of the approach are highlighted: Lean Production or Lean Management, Agile Product or Software development, and classical or plan- and hierarchy-oriented PM. In terms of their backgrounds, these management concepts can be associated with Japan, the USA and Germany. The cultural profiles of these countries are presented according to Hofstede's model and results.

Hofstede's dimensions

The term *culture* comes from Latin and in the relevant context of this book, means *civilisation* (synonym) or also *refinement of the mind*,[11] thus has a (historical) developmental aspect. Hofstede defines culture as the "collective programming" that distinguishes the members of one group or category of people from another.[12] Mental programming manifests itself in symbols, behavioural models, rituals and, last but not least, values.

---

8 see Hofstede 1997.
9 Hofstede 1997, p. 401.
10 Following Hofstede 1997, p. 5.
11 cf. duden.de: "Culture".
12 see Hofstede 1997, pp. 401–404.

Hofstede has been conducting worldwide studies at IBM since the 1960s. He structured the empirical results obtained in this way into the five dimensions that ultimately formed the classification system of his cultural patterns:[13]

Power distance
Power distance is the degree to which the less powerful members of an organisation accept or even expect the unequal distribution of power.

Individualism vs. collectivism
Individualism represents a form of society in which social ties beyond one's own family are not very strong. People take care of themselves and their immediate family first. In collectivism, on the other hand, people live in extended groups with a strong cohesion. There is an unquestionable loyalty to the group, which is rewarded with lifelong protection.

Uncertainty avoidance
The degree to which members of a culture feel threatened by uncertain or unknown situations is described by the cultural dimension of uncertainty avoidance.

Femininity vs. masculinity
By femininity, Hofstede refers to a society in which gender roles overlap. Men and women are equally considered modest, sensitive and concerned about quality of life. In masculine societies, on the other hand, social gender roles are clearly defined and differentiated. Men are supposed to be assertive, tough and materially oriented – women modest, tender and focused on quality of life.

Short-term vs. long-term orientation
This dimension is also called *Confucian Dynamics*. Short-term orientation means cultivating values related to the past and present, such as preserving traditions and face, and fulfilling social duties. The opposite stands for orientation towards virtues that are oriented towards future success, such as thrift and perseverance.[14]

In order to quantify the characterisation of a culture according to the cultural dimensions, Hofstede developed dimension-specific indices, e.g. the power distance index, which were determined with the help of specific questions and Likert scales. In this way, Hofstede assigns five indices to each (national) culture, which were determined on a country-specific summary basis using the IBM-related studies. High values (which are not standardised, but de facto distributed from about 1 to 100) describe a great power distance (i.e. acceptance of power), great individualism, strong masculinity, strong avoidance of uncertainties as well as a long-term orientation according to their definitions.

Of course, it should be noted that Hofstede's studies are already about 50 years old and that the basic population consisted of IBM employees, which is sometimes criticised. Hofstede turns these objections into the opposite, since the otherwise rather homogeneous population

---

13 see Hofstede 1997, p. 402.

14 This cultural dimension is not based on the IBM studies, but on the so-called *Chinese Values Study*, which was conducted among students in 23 countries, see Hofstede 1997, p. 226 f.

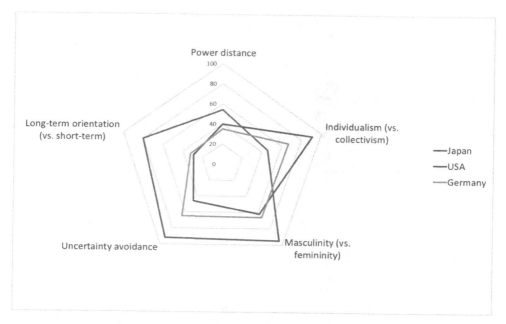

FIGURE 7.3   Country profiles according to Hofstede.

(e.g. with regard to the occupational profile of several hundred thousand worldwide employees) would even bring out the different characteristics more clearly. Therefore, the Hofstede model will be used as a guiding scheme in the sense of an orientation aid. Basically, it should be emphasised once again that beyond cultural imprinting, there are considerable differences between individuals within a country culture, which, even according to Hofstede, can be greater than the differences between the country cultures themselves. However, we can use the country values because they are based on a large, representative number and on the fact that most people are strongly influenced by social control.[15]

Profiles of the countries and management systems

Hofstede's analyses have shown the results in Figure 7.3 for the three countries under consideration.[16]

There are clear cultural differences between the countries, which can be quantified with the help of Hofstede's indices and dimensions. Maximum differences can be seen in the dimensions of long-term orientation and uncertainty avoidance, where Japan is clearly the most pronounced, and individualism, which clearly dominates in the USA.

In the following, I will attempt to classify the management methods Lean Management, agile development and classic PM according to the same scheme and compare them with the different countries as examples – albeit in an associative way. The result is shown in Figure 7.4.

15 see Hofstede Insights n.d.
16 with data material from Hofstede 1997, p. 30 f., p. 69 f., p. 115 f., p. 157 f. as well as quoted from Lingnau et al. 2004, p. 17.

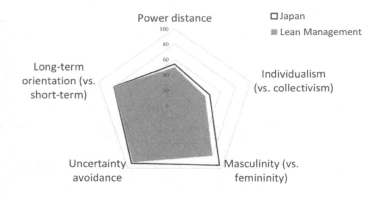

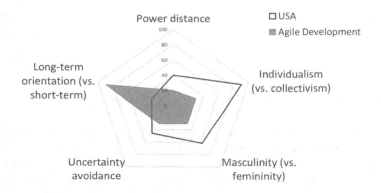

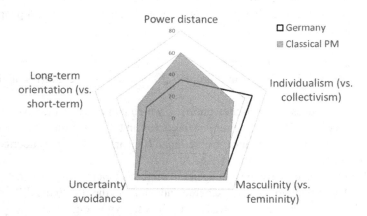

FIGURE 7.4 Comparison of cultural country and method profiles (I).

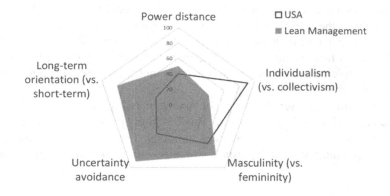

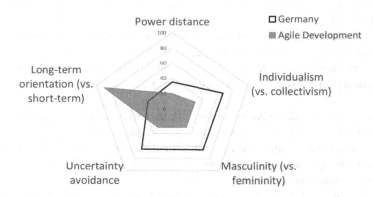

FIGURE 7.5   Comparison of cultural country and method profiles (II).

It must be emphasised once again that the classification of the methods in Hofstede's system only corresponds to an initial assessment in the sense of a thesis. The following scale was used for the mapping: Very low (10), relatively low (30), neutral (50), relatively high (70), very high (90).

Some interesting questions can be derived from this initial assessment. Is it a coincidence that …

- Lean Management, which is part of the Toyota Production System has a high correspondence with the cultural profile of Japan?
- agile approaches as an adaptation of Lean Management have a particularly strong development in the USA, which obviously demands significant cultural changes in the product development processes?
- the individualism that is particularly pronounced in the USA is curbed by agile methods?
- there is a high level of cultural agreement in Germany with the values that correlate with plannability?

My theses on these questions are that this is no coincidence, which could give rise to further scientific investigation. If this is the case, then it means that the origin and cultural conformity of methods must be taken into account if they are to be used in different cultural profiles. Figure 7.4 indicates that, on the basis of a fundamentally good fit, there are elements that can be used by a (new) approach and are supplemented by an improvement in other areas. The illustrations in Figure 7.5 can also give insight here.

The two graphs each show a greater deviation for management approaches that have been imported into the respective society, as it were. This definitely provides an indication of the difficulties to be expected when introducing methods. Lean PM as a – partly situational–symbiosis of agile and plan-based approaches requires in particular a corresponding analysis.

Behavioural typing according to Lewis

Moving more to the level of individuals, Lewis can give indications with his behaviour-based classification into linear-active, multi-active as well as reactive people. His typology includes the types shown in Figure 7.6.[17]

Lewis also assigns these characteristics to countries where they are dominant, but also emphasises that there are, for example, typical characteristics associated with professions. For example, a salesperson requires a different profile than a development engineer. The aim here is therefore more to characterise the people who (should) work in a project or a PM system. In summary:[18]

- Linear-active people are task-oriented, highly organised planners who complete action chains by doing one thing at a time, preferably in accordance with a linear agenda.
- Multi-active people are emotional, communicative and impulsive people who attach great importance to family, feelings, relationships and people in general. They like to do many things at once and are poor followers of agendas.
- Reactive people are good listeners who rarely initiate action or discussion, preferring to listen first and establish the other person's point of view, then react to it and form their own opinion.

But when are the appropriate people needed and well deployed? Here are some considerations:

Type:	Required skills:
Linear Active	organise, plan, see problems, analyse consequences, follow consistent guidelines, access rational thinking, generate data and challenge objectively
Multi-active	generate enthusiasm, motivate, convince, create a positive social atmosphere, gain access to emotions, promote dialogue and challenge personally
Reactive	harmonise, act intuitively, be patient and see the Big Picture, think and act long-term, access feelings, listen, empathise and challenge holistically.

17 cf. Lewis 2006, quoted from CrossCulture n.d.
18 see CrossCulture n.d.

Linear-active	Multi-active	Reactive
speaks half the time	speaks most of the time	listens most of the time
does one thing at a time	does several things at once	reacts to the action of the partner
plans step by step forward	plans only the overall outline	considers general principles
polite but correct	emotional	polite, indirect
partly conceals feelings	shows feelings	hides feelings
confronts with logic	confronts emotionally	never confronts
does not like to lose face	has good excuses	must not lose face
rarely interrupts	frequently interrupts	does not interrupt
job-oriented	people-oriented	very people-oriented
sticks to facts	feelings before facts	declarations are promises
truth before diplomacy	flexible truth	diplomacy over truth
sometimes impatient	impatient	patient
limited body language	unlimited body language	subtle body language
respects bureaucracy	locates the key person	uses connections
separates the social and the professional	mixes the social and the professional	combines the social and the professional

FIGURE 7.6 The characteristics of cultural typification according to Lewis.

The connection with the use of agile or plan-driven or hybrid approaches becomes immediately clear. Those who are not "polarised" towards planning will not create a good plan, and those who cannot live well with uncertainties will have difficulties with continuous reprioritisation in relatively short time intervals.

As a consequence of these considerations, corresponding measures of change management, and further and advanced training must be carried out, especially for those who have difficulties in adapting to the part of the typology. It is in the nature of human beings that this can only happen to a limited extent, for example by strengthening knowledge and confidence. Ultimately, the question is whether or not the person can participate in further development. This has already led to non-employment and dismissals in companies or to "bad teams" in Scrum projects, in which employees unsuitable for the method were "parked" harmlessly (e.g. by having them carry out only routine operational work).

Time orientation according to Hall
The way people deal with time is a relevant component in many of the known cultural classification schemes. Since plan-driven and agile approaches differ precisely in this characteristic, it is worth taking a closer look here. For this purpose, Hall's cultural category "monochronic – polychronic" will be examined. The pair of terms is derived from the Greek or Latin *chronos* (time) and *mono* (one) as well as *poly* (many or much) and means something like: What do I do with time? Do I divide it up and systematically do one thing after another – as planned and structured as possible? Or do I divide it up and do many things at the same time according to a loose system that is always open to surprises?[19] Hall divided people according to the following behaviour patterns:[20]

19 see Zaninelli 1995.
20 cf. Hall 1990, pp. 1–20.

Monochrony	Polychrony
Do one task at a time	Do many tasks at the same time (multitasking)
High concentration	High distraction
Appointments are taken seriously	Dates have no meaning
Orientation to plans	Plans have no meaning
Disturbance of others is avoided	Disturbance of others is accepted
High punctuality	Low punctuality
Methodological work	Patience is easily lost

This is illustrated by the example of the course of meetings in Germany (monochronic) and France (polychronic). Punctuality is very important at meetings in Germany. Being late is an affront to the other participants and means unreliability. The agenda is worked through point by point in a sequential manner. In France, meetings are more informal. It is mainly about personal exchange of opinions and information. No one is upset if they are late and the agenda is seen as a dynamically adaptable timetable.[21] The transfer to the affinity for plan-driven, milestone-oriented versus flexible, adaptive procedures is obvious. Last but not least, the definition of procedures such as in Scrum which is systematically observed in terms of subject matter and time (Scrum Master), makes sense for both behaviour patterns. Polychronic people thus get a framework of order, monochronic people more flexibility.

7.1.2 Systemic view

In Lean Thinking, great importance is attached to the question of cooperation between the participants. This can be divided into a horizontal and a vertical view: In addition to the requirement of integrating the process customers into the development of the results intended for them (horizontal view), the question of how leadership should be designed in the project – i.e. which function the project management has and which the project team has (vertical view) – must also be answered. This chapter looks at the vertically oriented question.

The St. Gallen Management Model

The stereotypical answer to the question of leadership is "Self-organisation of the teams!" in agile and "Hierarchical structure!" in the classical model. The latter is expressed by the traditional role of the project leader who manages the team in all aspects. The former (strongly influenced by Scrum) is expressed in many cases by the more moderating, advisory and empowering role of the Scrum Master and the requirements-, product- and benefit-oriented role of the Product Owner.

The perceptions of team management therefore differ significantly in both approaches. Due to the inherent contradictions, a mixture is theoretically and practically difficult or even

21 see Untereiner 2013, pp. 177–179.

impossible. The question therefore arises as to how the right approach can be found for a concrete project. For this purpose, let us systematically look at the term *management*. Following the renowned *St. Gallen Management Model* (SGMM), it makes sense to distinguish between *normative, strategic* and *operational* management.[22]

Normative means serving as a guideline or norm, as well as representing a rule, a standard for something.[23] In the SGMM, normative management processes serve to clarify the normative foundations of entrepreneurial activity, which includes, for example, the establishment of fundamental principles of conduct for dealing with stakeholder groups or with risky technology. With regard to project management, the establishment of framework conditions for the implementation of projects can be defined in this context. These should be in line with legal and regulatory requirements (e.g. data protection), professional standards that are common and applicable in the industry (e.g. norms) and also with the company's mission statement (e.g. sustainability). This high-level view thus refers primarily to the organisation's environment. Stakeholders are the legislator, the shareholders and the market participants. However, normative guidelines also arise internally within the company, which can be seen in the area of quality management as an example of internal work guidelines. Normative management is usually institutionalised through contextually authorised representatives or staff units (e.g. occupational safety representatives).

According to SGMM, the term *strategic* refers to the competitive, long-term safeguarding of the company's future by building sustainable competitive advantages. Tasks include, among others, the establishment of strategic cooperations and, last but not least, capability development through process design projects. This task is typically the responsibility of the organisation's senior management and is operationalised, i.e. broken down, to the middle management executives.

The SGMM defines *operational management* primarily as the process management of the individual service provision and support processes. Accordingly, the tasks of operational process management include the situational regulation of day-to-day business issues that are not already structurally regulated and thus decided in advance, the prioritisation and targeted allocation of resources to orders (scheduling) as well as ensuring the quality and continuous optimisation of the process. The SGMM also counts ensuring constructive cooperation and goal-oriented behavioural influence as operational management at team level (staff leadership). With regard to financial management, reporting and controlling of the performance and impact of management decisions are also included.

Equipped with this thought model of the differentiated consideration of management tasks, the question posed at the beginning about the right approach for the design of leadership in the project can be answered structurally in principle: Only operational management in the project is suitable for self-organisation. All the elements of operational management identified by the SGMM can be found in the role and procedure model of Scrum, for example. However,

---

22 in the version of the SGMM of 2005, see Rüegg-Stürm 2005, p. 70 ff.
23 cf. duden.de "normativ".

self-organisation should not be confused with strategic and normative management. This is always to be exercised from a higher, correspondingly responsible position (strategic view) or from outside (normative regulative). Both dimensions ultimately provide the guard rails within which the team can manage itself operationally. This is also taken into account, for example, in large, complex projects by establishing roles such as the leading system architect or similar, who exercise a corresponding directive competence.

But here, too, practical experience shows:

> Letting teams work out processes in a self-organised way, which can basically also be clearly predefined, is a waste!

In Section 2.3.5 the *cross of complexity* was introduced. According to this, projects can be classified as simple, complicated, dynamic or complex systems. Depending on the characterisation, different norm strategies result. The characterisation also has implications with regard to the question of the design of leadership:

The project is characterised as ...	Operational management ...
Simple	... is hardly necessary, can be reduced to a minimum; depending on the constellation, self-organised or hierarchical possible
Dynamic	... should be self-organised and rule-based, using the competence of the team; higher-level leadership becomes a bottleneck
Complicated	... must be strengthened; higher-level directorate increases efficiency
Complex	... should be delegated downwards as far as possible; strategic management as a higher-level management task must be exercised more strongly

Overall, large projects with many substructures (subprojects or similar) can be structured according to the described dimensions of management. The question should be asked for each unit: Who is responsible for normative, strategic and operational management? In this way, the network of hierarchical structures, roles and *cross-sectional functions* (quality management, etc.) can be designed in a targeted manner.

7.1.3 Employee-related leadership

With the considerations derived from the SGMM on the nature of management tasks in projects, a generic, structural approach is available. As a project-related application of the universal SGMM, which has been continuously developed since the 1970s, and from my own application experience in large-scale projects, the approach is evident as a practice-oriented guideline for organisational project design. Nevertheless, the concept leaves out one crucial aspect – the

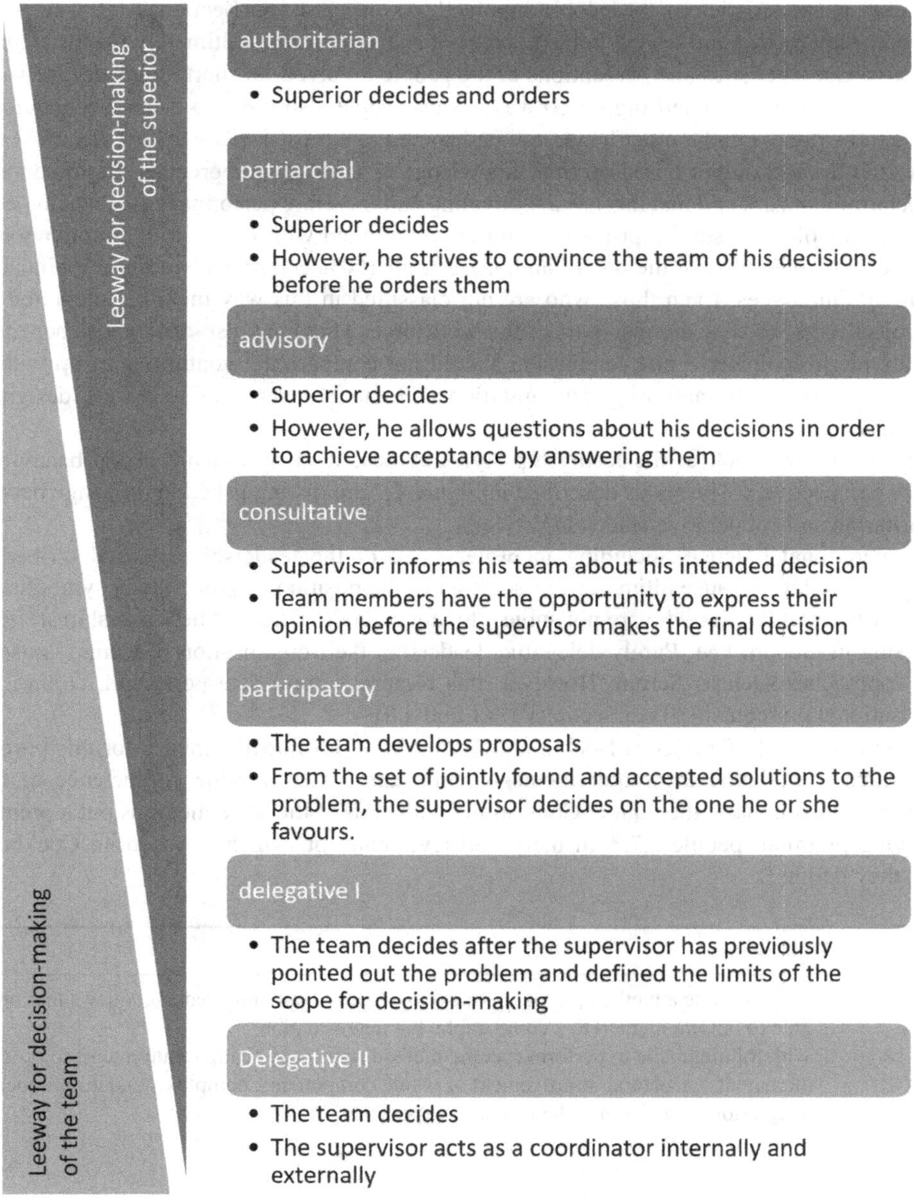

authoritarian

- Superior decides and orders

patriarchal

- Superior decides
- However, he strives to convince the team of his decisions before he orders them

advisory

- Superior decides
- However, he allows questions about his decisions in order to achieve acceptance by answering them

consultative

- Supervisor informs his team about his intended decision
- Team members have the opportunity to express their opinion before the supervisor makes the final decision

participatory

- The team develops proposals
- From the set of jointly found and accepted solutions to the problem, the supervisor decides on the one he or she favours.

delegative I

- The team decides after the supervisor has previously pointed out the problem and defined the limits of the scope for decision-making

Delegative II

- The team decides
- The supervisor acts as a coordinator internally and externally

Leeway for decision-making of the superior

Leeway for decision-making of the team

FIGURE 7.7   Leadership styles scale.

individual characteristics of the people involved. As discussed in Section 7.1.1 what seems good and right or bad and wrong in the context of social structures ultimately results from the subjective characteristics and perceptions of the people involved. In short, not everyone is able and willing to work in a self-organised way; many people also prefer a clear specification of what is to be done. On the other hand, it is undisputed that most *brain workers,* i.e. those who deliver the desired output based on their knowledge or creativity, appreciate the freedoms of self-determined work and that this has a motivating and therefore performance-enhancing effect on them. Sociological studies prove this and are often cited with regard to the motivation for agile working methods. On the other hand, there is no 100% congruence of these typifications among all employees. Even those who are not classified in this way must be integrated into the project work, as they are also part of the workforce. There are also differences depending on the type of company – one cannot and should not compare the youthful start-up with the long-established public authority. The ambition of Lean PM demands universal, designable solutions here.

First of all, we look at the leadership styles as described by Staehle in his behavioural science perspective.[24] The styles described in Figure 7.7 are distinguished in the range between authoritarian and cooperative leadership styles.

In operational practice, including in projects, all of the leadership styles described are represented today, because ultimately, in addition to the position of power, they symbolise the character traits of leaders who are not subject to any zeitgeist. The middle leadership styles are promising in modern PM. Purely delegative leadership (self-organisation of teams) is used in agile approaches such as Scrum. However, this requires appropriate personnel, cultural and organisational prerequisites (see Sections 7.1.1 and 4.8).

In the context of IT projects, Boehm and Turner have developed a metric for this purpose, with which they define the requirements of agile approaches for the competence of team members.[25] In summary, they quote Constantine with "All of the agile methods put a premium on having premium people ...".[26] In their skill-level concept, which is based on Cockburn's work, they define:[27]

Level:	Employee is ...
Level 3	... able to revise a method (break its rules) to adapt it to an unprecedented new situation.
Level 2	... able to tailor a method to a comparable, but new situation.
Level 1A	... with training, able to perform specific method steps (e.g. sizing requirements to product increments, structural improvement of whole components, complex standard product integration). Can become level 2 with experience.

24 see Staehle et al. 2014, p. 328 ff.
25 see Boehm/Turner 2009, pp. 46–49.
26 see Constantine 2001.
27 see Cockburn 2002.

Level:	Employee is ...
Level 1B	... with training, able to perform procedural method steps (e.g. coding a simple method, simple structural improvement, following development standards and change management procedures, performing tests). With experience, can master some level 1A skills.
Level -1	... may have technical skills but is unable or unwilling to collaborate or follow methods of joint working.

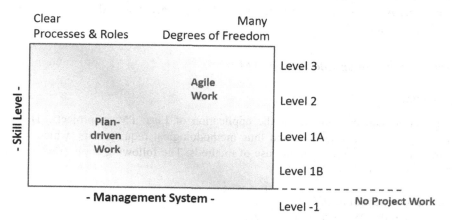

FIGURE 7.8   How are the available employees to be classified?

This list has been freed from the (inconsequential) IT specifics of the original, since it does not seem necessary to restrict the significance to pure IT projects. Boehm and Turner state, *mutatis mutandis*, that even the plan-driven methods naturally cope better with competent people but are generally better able to plan the project and produce the output in such a way that even less capable people can contribute. With regard to their use in projects, they rate: Employees of ...

- Level -1 should be quickly identified and used for tasks other than working in agile or scheduled teams.
- Level 1B require extensive guidance and work well in plan-oriented environments.
- Level 1A need guidance but can work well in agile teams.
- Level 2 can manage manageable projects independently but need guidance in unpredictable constellations.

Figure 7.8 summarises Boehm and Turner's findings in an own portfolio representation, which can be used to classify the employees available for a project.

Depending on the complexity of the project or solution, Boehm and Turner conclude that agile projects need at least 30% full-time Level 2 and Level 3 staff and cannot use Level 1B and Level -1 staff. Meanwhile, plan-driven projects initially require 50% Level 3 staff, which can be reduced to 10% as the project progresses; generally, up to 30% Level 1B staff is practicable, but again, no Level-1 staff.

In many cases, companies cannot, in practice, set themselves up in accordance with what is considered to be the ideal level of staffing. The approach chosen must therefore be based on the staff available. A manager's leadership style should not be based on his or her preferences, but on the individual needs of the members of his or her team. One needs more guidance and planning security, the other more degrees of freedom to develop their potential. This flexibility of situational adaptation is what distinguishes a good project manager.

Applied to the introduction of Lean PM, this means taking into account the people involved in addition to all the technical aspects: What are they able to do and what do they want to do in the project? A corresponding criterion was taken into account in the conception of the Agilometer (see Section 4.8).

7.1.4 Project-related application of Lean PM

7.1.4.1 Derivation

Different scenarios can be defined for the application of Lean PM in projects. These are characterised by different objectives and thus methodological requirements, which leads to specific procedures and a corresponding use of methods. The following application scenarios can be identified in particular:

- General improvement of the CPM system,
- a priori set-up of the PM system of a project,
- problem analysis and elimination in a concrete, ongoing project
- and last but not least, Lean Thinking as a mental guideline in the execution of day-to-day project work.

The latter is certainly the least formalisable process, as it simply means the non-formally anchored application of Lean principles and methods based on the competence of individual project managers and teams. As the examples in Chapter 5 show, Lean elements can and should be applied situationally when operationally required in the project. This means that in designing the project processes and structures for the best possible execution, those involved are guided by Lean Thinking and use these techniques. In the long run, this changes the project culture of the organisation through the expected success – if you will, through the power of the factual.

This positively connoted guerrilla tactic, i.e. the bottom-up approach, can of course also give way to a more open and targeted, official path – namely the introduction of Lean PM as a project or change measure.

7.1.4.2 Improvement of the corporate PM system

This scenario is about the sustainable anchoring of Lean PM as the organisation's corporate PM system. or its further development. Typically, this leads to the adoption of a PM manual for the organisation and the establishment of corresponding processes, regulations, roles and structures (project management offices, etc.). It is, so to speak, the "big shot" in the direction of Lean PM and carries the danger of a head-birth. The advantages lie in the clear positioning. Probably many organisations know the problem of management system guidelines packed in manuals (or intranet pages), which are perceived as bureaucracy by most employees and are not followed. Lean Thinking should be the guiding principle when introducing Lean PM to avoid this effect. Customer and benefit orientation while avoiding waste are the manifest, obvious evidence of this.

In order to proceed in a targeted manner, it is advisable to conduct a maturity analysis of the organisation's PM. In general, well-known maturity models can be used here, e.g. the *Capability Maturity Model Integration, Project Management Maturity Model* or proprietary models such as the *Organizational Project Management Maturity Model (OPM3)*. Following Kerzner, organisations have reached a high level of PM maturity when they have established methods and processes that offer a high probability of repeated success. Accordingly, organisations achieve excellence when they have reached a level of maturity that enables a continuous stream of successful projects, measured in terms of benefit for customers and the company.[28]

Typical ingredients for conducting a PM maturity analysis are a graded assessment scheme and a PM reference model against which the assessment is made. The following sequence is suitable as a clearly graded assessment scheme: Processes ...

[0] ... are not present,
[1] ... are carried out individually in each case,
[2] ... are partially defined for the organisation,
[3] ... are fully standardised and established,
[4] ... are continuously developed and
[5] ... are systematically measured and benchmarked.

The Unified Project Management Framework is suitable as a procedural reference model (see Section 1.1.2) or the well-known models of the various PM organisations (PMI, DIN/ISO etc.). Depending on the orientation of the targeted system, project portfolio management is also relevant, e.g. when it comes to the question of initiating and approving a project. With the help of the reference model, the maturity level can then be examined process by process, role by role. An actual profile of the organisation is created, which systematically shows the current positioning. Based on the definition of a target state, for example oriented to the requirements of the competition, the fields of action can be derived and prioritised. In the subsequent derivation of measures, methods such as Target Value Design, the Ishikawa diagram (see Chapter 4) and others should be used.

7.1.4.3 Setting up the PM system of a project

The design of the PM system of a concrete project has a different focus. Here, the focus is not on organisational PM, but on an individual project with its specific framework conditions. Thus, it is not a PM manual of the organisation that is created, but a project charter of the project at hand. The project manual is the meta-documentation for PM (processes, roles, regulations, methods

---

28 see Kerzner 2003, p. 52, pp. 657–661 (PMMM); SEI 2009 (CMMI); PMI 2005 (OPM3); cf. Wendler 2009, pp. 26–39.

and tools, organisational structure, scope, boundary conditions, etc.). Synonyms of the project manual are project management plan (according to PMI) or project initiation documentation (according to PRINCE2).

In the project-specific design of the PM system, the question is answered: Which PM practices do we apply in the project so that the framework conditions and objectives of the project are met in the best possible way and thus the success of the project is promoted? This scenario is called an *adaptation* of a project's *PM system* and is described in Chapter 5. It is a central element of operational Lean PM because it follows the realisation that the uniqueness of a project always requires an individual, tailor-made solution. This is not a contradiction to establishing Lean PM at the level of the corporate PM system. On the contrary, the demand for project-specific design of the PM system should be a central component of the Lean Corporate PM system.

7.1.4.4 Problem solving in an ongoing project

The scenario depicted above describes the *a priori* design of the single-project PM system. But even in an ongoing project, it may be necessary to analyse the use of PM practices in order to find more adequate practices. On the one hand, this requires the Lean principle of continuous improvement. Projects always generate new knowledge in their life cycle that needs to be implemented. Regular retrospectives ensure the systematic identification and handling of Lessons Learned.. On the other hand, projects often get into difficult waters – crisis situations in which it is needed to analyse and eliminate problems and their causes. This happens, for example, in the form of extraordinary reviews of the project and the PM. In a project to develop a new technical product platform, for example, the project manager reported that they deliberately adopted an agile Scrum approach and had created the basic formal framework conditions, such as co-working in a project war room. But that the project was nevertheless threatening to fail because the collaboration was not working. In such a situation, a review becomes necessary in order to identify improvement measures.

Similar to the design of the CPM system, a maturity analysis based on a PM process reference model can also be carried out when reviewing an individual project. The guiding questions oriented towards Lean PM are, for example:

Customer	Who are the project customers for the project results?
	Who are the process customers for the activities under consideration?
	What is important for them? Etc.
Value creation	How is value added in the process under consideration?
	Where are the activities to be characterised as waste? Etc.
River	Where does the flow of information get bogged down?
	Where are decisions delayed? Etc.
Pull	At which point in the processes is pre-production (of results) e.g. not expedient due to expected dynamics?
	Where has overload arisen due to harmful multitasking? Etc.
Learning	How can we ensure that relevant knowledge is always available?
	How and when should we learn from experience?

7.1.4.5 Use in the project

In order to take the Lean idea into account in the life cycle of a project, some elements should be considered in the respective phases of execution. Depending on the context, basically all known practices can be used, and some of the presented practices (see Chapter 4) are worth highlighting. If one transfers and generalises the empirically gained knowledge of Lean Project Delivery in infrastructure projects (Lean Construction), the following indications emerge:[29]

Project initialisation and preparation

Structure the project contractually and organisationally to enable Lean PM. That is:

- Use relational contracts and provide for cross-functional teams.
- Bring benefit expectation, possibilities and boundary conditions in line.
- Set targets for project scope and costs based on agreed benefit expectations, opportunities and constraints.
- Set the framework for experimentation and learning.

Project planning and operationalisation

- Pursue a design strategy according to Set-based Design, i.e. create alternatives and keep them open as long as possible.
- Align the design with the agreed costs, taking into account the customer value (Target Value Design).
- Choose a project approach that is coherent with the framework conditions for the development of results – including realisation, solution integration and pre-transfer of operations.
- Integrate product and system specifications as well as installation and operating instructions throughout in an integrated, database-supported IT solution.

Execution

- Make project processes predictable through goal-oriented and efficient project controlling.
- Apply appropriate Lean practices in the realisation, e.g. 5S, Value Stream Mapping, Failure Mode and Effects Analysis, Poka Yoke, Gemba, cause-effect analysis, modular development, etc.
- Make conceptual decisions at the Last Responsible Moment to reduce the risk of design changes.
- Realise based on defined releases, using the MuSCoW rule.
- Conduct first-run studies to improve safety, quality, time and cost of operations – and thereby ...
- ... Involve the project team in designing, testing and improving the way of working.

---

29 cf. Ballard et al. 2007, p. 145 ff.

- Achieve quality through planned preparation and systematic identification, rapid correction and prevention of errors.
- Get feedback on project management effectiveness and suggestions for improvement by interviewing developers, service providers and other stakeholders.

Transfer to business and operations

- Use commissioning and going live to check compliance with the requirements.
- Transfer information (models, actual configuration, system manuals) to users and the operating organisation for use in operation and maintenance.
- After the usage has been swung in, carry out an evaluation to check the appropriateness of the concept and solution.
- Collect feedback from members of the project team and other stakeholders on the lessons learnt.

First and foremost (but not only) among the important stakeholders to be included in the life cycle of the project according to Lean PM systematics are the project customers. Every project benefits from competent customer representatives who are available as comprehensively as possible in terms of time and at the location of the event. This sounds simple, but it places demands on their involvement that cannot always be met.

In this context, Boehm and Turner have studied more than 100 projects that have produced e-services (i.e. services delivered over the Internet using information and communication technology) and from this they have formulated the beautiful acronym of the *CRACK client representative*.[30] Those representatives of the customer who are to be meaningfully involved in the project should therefore be CRACKs. CRACK stands for Collaborative, Representative, Authorised, Committed and Knowledgeable. The following associations apply:

Are the customer representatives not	*they will ...*
cooperative	... sow disunity and frustration, leading to loss of team morale.
representative	... make the project staff deliver unacceptable products.
authorised	... accept delays in approvals and acceptances or mislead the project through unauthorised commitments.
committed	... not provide the necessary input and not be there when the project team needs them most.
knowledgeable	... cause delays and unacceptable, defective products.

In practice, it is not uncommon to observe that departments do not send their CRACKs into the project to participate in the conception, development and testing of the solution, but rather, for example, those who can best be dispensed with in the company at the moment. The

---

30 see Boehm/Turner 2009, pp. 44–45.

aforementioned consequences of waste are then almost pre-programmed. For this reason, it should be clarified in advance when introducing technical requirements into the project who on the part of the customer, i.e. the *specialist department,* will represent them at this point – and whether they fulfil the CRACK requirements. In the worst case, in a project in which I was involved, a corresponding miscommissioning led to a complete reworking of a technical concept – after it had been submitted on time in the first, supposedly agreed version. Another typical effect is measurable by the error rate in final tests, where it only becomes clear that essential elements were not implemented according to requirements.

7.2   Procedure for the introduction

7.2.1   Introduction of Lean PM as a change project

Changing a management system in a company is first and foremost an organisational project and thus inevitably a change project. This of course particularly concerns the top-down approach, i.e. in the case of Lean PM the redesign of the CPM system. Changes in the behaviour of employees and the culture of the company, such as the communication or error culture are pursued.[31] The introduction of Lean PM therefore usually means cross-departmental, sometimes radical and comprehensive changes. Resistance and unrealistic expectations regarding the results must be expected from those affected. Challenges with regard to cultural changes through Lean PM can be exemplified by the 3Gs for Lean PM:

Participation	Customer involvement, delegation of responsibility to the team, transparency, etc.
Pareto principle	Openness to solutions, black boxes in planning and scope etc.
Fit	Flexible adaptation of methods, processes as well as roles and leadership etc.

In this respect, the insights and methods of change management should be incorporated. The task of change management is to achieve an optimal design of organisational change processes.[32] Resistance from employees, inadequate process control, too fast a pace of change as well as unclear objectives are – in this order – often decisive causes for the failure of change projects.[33] The following rule of thumb can be used to gain a basic understanding of the factors that influence overcoming resistance and the energy needed to maintain the status quo:[34]

	Dissatisfaction with the current state
times	*Vision/attractiveness of the target state*
times	*Measures/concrete implementation steps/achievable first successes*
should be bigger than	*Resistance to change/energy to preserve what exists*

---

31 cf. Kuster et al. 2019, p. 19.
32 see Lauer 2019, p. 3.
33 see Schott/Wick 2005, p. 196.
34 see Kuster et al. 2019, p. 19.

There are different approaches to implementing change projects. Many models try to provide a framework for action by dividing the change process into phases and describing typical phase observations. In all models, three overarching phases can be identified that are passed through in a successful change process:[35]

- Initialisation of change
- Transformation of the system
- Consolidation and stabilisation of change

Basically, there are two directions from which change is initiated, although in reality there are often mixed forms: Change can be initiated from above (top-down) by management, which takes over the planning of the change process and the formulation of vision and goals. Or it can be initiated from below (bottom-up) through staff insights into undesirable developments and potential for optimisation. Since employees are instructed to change in the top-down variant, more resistance must be expected here and there may be excessive expectations at the management level. The bottom-up variant has the advantage that the change is borne by the experts on the ground, as they identify the necessary restructuring. However, this can lead to targets being set too low and only limited visions being developed. In addition, there is usually a lack of competence to steer the change process.[36]

7.2.1.1 Procedure model

If Lean PM is to be established as a CPM system the well-known 8-step model by Kotter can be applied. The first four stages, as shown in Figure 7.9, help to thaw the hardened status quo in order to pave the way for the subsequent introduction of the new methods in stages 5 to 7. Finally, the last stage anchors Lean PM in the corporate culture and thus ensures the success of the application and the sustainability of the change.

According to Kotter, a distinction must be made between *management* and *leadership* in the context of change, with the latter being responsible for 70–90% of a successful transformation. In contrast to management, which tries to efficiently operate the company system consisting of technology and people through optimised processes, it is the responsibility of leadership to shape the company. In doing so, a vision for the future and corresponding implementation strategies should be developed as a guide, on the basis of which communication is then built to align employees. This enables the development of coalitions and teams in which the goals and approach have been understood and accepted. Leadership also has the task of motivating and inspiring those involved by taking into account the fulfilment of human needs and thus empowering those involved to break down the barriers to change.[37]

---

35 cf. Kotter 2011; Krüger/Bach 2014.
36 see Krüger/Bach 2014, pp. 56–59.
37 see Kotter 2011, p. 22.

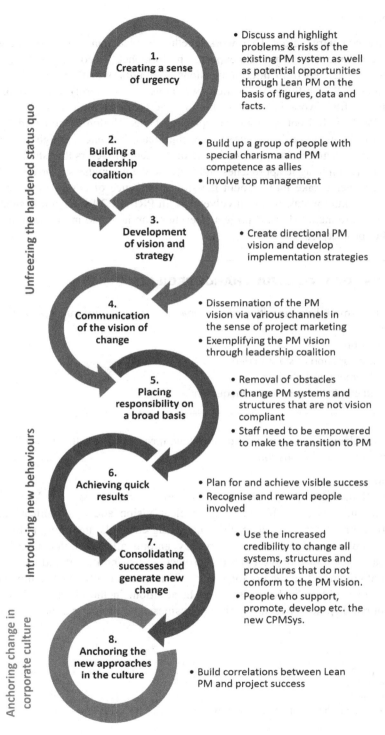

FIGURE 7.9   The 8-step model according to Kotter.

Own illustration according to Kotter 2011, p. 18.

7.2.1.2 Success factors

In projects, people are the central actors. They work together as a team and communicate with each other to achieve change and deliver the expected performance. According to Sprenger, employee performance is formed from three dimensions: Willingness to Perform, Ability to Perform and Performance Possibility.[38] Motivation (willingness to perform) is an important component. Appropriate framework conditions should be created in project work so that employees can develop their full potential and acceptance of the solution is increased.[39] The form of delegation of tasks in the project context has a significant influence. A high degree of self-determination in the design of requirements is one of the central factors that shall lead to increased intrinsic motivation in agile approaches and increase the willingness to adapt.[40] Both classical and agile approaches take into account the strong influence of communication and include appropriate elements for stakeholder involvement. In PRINCE2 for example, supplier and user representatives are included in the project steering committee and in Scrum users are an integral part of the (project) team.

SUCCESS FACTORS OF SUCCESSFUL CHANGE PROJECTS

- Communication (form of dialogue with project management and senior management)
- Vision development
- Learning organisation
- Active role and participation of stakeholders
- Leadership development and training
- Commitment and credibility of senior management

In summary, successful change management should initiate measures in the three directions of *information, empowerment* and *mobilisation.*[41]

Information – What?
Targeted communication, planned and structured by project marketing, should provide broad-based information. It is advisable to create a clear vision and communicate clear goals. Communication through the various project marketing channels, including individual discussions, should aim to create a sense of urgency among all stakeholders and thus systematically increase acceptance of the Lean PM project. The commitment and credibility of senior management are important factors for success. This makes it all the more important to build a management coalition in Kotter's sense. In addition, in the absence of project acceptance, the resource dispute between day-to-day business and project can hardly end in favour of the project.

---

38 see Sprenger 2014, p. 183 ff.
39 see Kuster et al. 2019, p. 275.
40 see Goll/Hommel 2015, p. 12 ff.
41 see Fleig 2019, p. 2; Capgemini 2006, p. 47; Sterrer 2014, p. 139.

Empowerment – How?

To enable empowerment, the team members and those affected should be given the skills to implement and live the changes through training measures, clear role descriptions, process descriptions, etc. The implementation of a learning organisation and the granting of freedom in the provision of services contribute to the development of productivity and increase identification with the project and its results.

Mobilisation – What for?

As early as possible and holistically, affected employees should be involved in the project and thus responsibility should be placed on a broad basis. Through large group methods, such as facilitation, those involved and affected should be given the opportunity to participate in order to help with shaping the introduction of Lean PM and to find the meaning of Lean Thinking for everyone individually. Short-term successes should be planned for and their results published through the communication channels.

7.2.2   Technical procedure

Lean is a journey, not a destination.[42]

In the following section we look at the introduction of Lean PM as a project to change the organisation's CPM system. In doing so, the described requirements of change management are integrated – without repeating the already described elements one-to-one again. Depending on the context of the project, these are as described in Section 7.2.1. Figure 7.10 shows a possible implementation model for Lean PM in the organisation. It is based on a scenario in which a concerned group or individual who is convinced of Lean PM triggers and initiates the corresponding project.

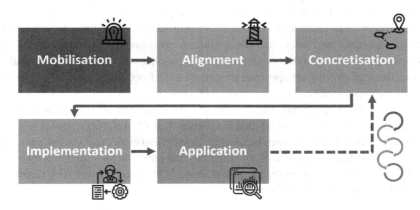

FIGURE 7.10   Process model for the introduction of Lean PM – phases.

---

42 Ballard et al. 2007, p. 136.

Due to its nature as a change project, the introduction starts with a mobilisation phase. In this phase, the problems of the organisation are to be identified as well as the opportunities that arise with the application of Lean PM. A proven instrument for this is the SWOT analysis (Strengths, Weaknesses, Opportunities, Threats and Risks). This is used to assess the current situation of the organisation with regard to project management. Weaknesses in project execution and the resulting risks are identified. For example, it is found that newly developed applications often have too high error rates during final testing by users, which creates the risk of delays and additional costs due to necessary rework. These risks can have an external (market, competition, etc.) and/or an internal focus (costs, employee satisfaction, etc.). This is equally true for the opportunities that arise for the organisation through the implementation of Lean PM. In general, current strengths will also contribute to the realisation of opportunities. For example, one strength of the organisation could be that the company usually has very long, stable and confidential customer relationships. This provides an opportunity to make projects more participatory and to involve customers in the development of solutions, as they are likely to value transparency and be forgiving of mistakes.

The results of this analysis are important because they can be used to mobilise the stakeholders for the change project: Clients, sponsors and other managers, affected employees and finally the project team. Awareness of problems and opportunities is an important success factor for a successful change (see Section 7.2.1), as is the positioning of a clearly defined and named principal who stands for the project in senior management and who, as a project client, desires the expected benefits. If applicable, a proxy can be installed who supports the project as a sponsor in the operational process.

So with mobilisation it is clear: The organisation wants the change and the project client is clearly positioned. The alignment phase follows. Here the goals and framework conditions of the project are defined in the project mandate. With the project strategy, the approach for the project is agreed: Will there be a pilot, if so, for which area? An area should be chosen that explicitly supports the project and that can later act as a multiplier – professionally and communicatively. However, piloting can also take place thematically by first addressing a selected process area. This could be, for example, the process of integrating technical requirements. The project team should be recruited and trained in the required methods, etc. Finally, to create transparency and tolerance, it's important not to neglect communication of the results of the phase. Depending on the organisational circumstances, the project proposal in this phase first goes through the project approval procedure as defined in the context of project portfolio management, if applicable.

After the project has been approved, the content can be concretised. This is where the practices of Lean PM come into play, as described in Chapter 4. A PM reference model, e.g. the Unified Project Management Framework can be used to identify and evaluate the weak points in project execution. The so-called *PII indicator* developed by Belvedere et al. – the *Priority Index of Intervention* – can be used to evaluate the identified waste.[43] This is determined by means of a survey of stakeholders and calculated as follows:

---

43 see Belvedere et al. 2019, p. 418.

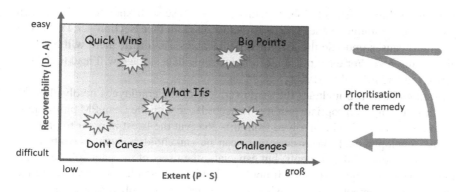

FIGURE 7.11   The Waste Heatmap for prioritising waste.

- For each type of waste identified, four questions are answered:
  o How often does this waste occur? (P = Probability)
  o How serious is this waste? (S = Severity)
  o How easily can this waste be detected? (D = Detectability)
  o How avoidable is this waste? (A = Avoidability)
- The answers, i.e. ratings, are given using a Likert scale, e.g. from 1 (very low) to 4 (very high).
- The index PII is then calculated as follows: $PII = P \times S \times D \times A$

The idea of the PII is an adaptation of the FMEA (Failure Mode and Effects Analysis, see Section 4.7.3). As with FMEA,[44] an index is calculated, which in this case ranks the types of waste, inducing a sensible prioritisation with regard to the further handling of the identified type of waste. The larger the PII, the more obvious the creation of a remedy is and should be addressed with priority (see Figure 7.11).

A heat map of waste can be constructed on the basis of the PII. The probability of occurrence (P) and the severity (S) quantify the *extent of* the waste, while the detectability (D) and the avoidability (A) jointly quantify its *remediability*. In this way, the "big points" can be identified, which have a large magnitude and are easy to eliminate. Likewise, other wastes can be characterised as "Quick Wins", "Challenges", "Don't Cares" and "What Ifs" according to their classification in the heat map. With the help of Lean PM principles and practices, solutions to avoid the identified wastes can now be found. The priority index of the intervention in turn helps to avoid waste by focusing on the essential items. The approaches to solutions lead to corresponding measures. In the form of a catalogue of measures, they are shared with the relevant stakeholders for the purpose of evaluation. Feedback from the stakeholders increases the maturity of the considerations and the implementation phase can begin.

The catalogue of measures is the (prioritised) backlog of the Lean PM implementation project. It is a living artefact that can always be enriched with new findings in the course of the further progress. In order to projectise the measures, they must be logically bundled in order

44 on the FMEA cf. DGQ 2012.

to realise synergies in the implementation. For this purpose, use well-known methods such as the 5-Why question technique or the (typed) Ishikawa diagram (see Section 4.7.4) to work out which (common) causes underlie the identified deficiencies. As a rule, there will not be a one-to-one relationship here. For example, a lack of electronic data storage will lead to delays, transmission errors and additional work, etc.

The concretisation phase also includes the empowerment of the employees involved in the solution, for example through appropriate training. In accordance with the Lean PM philosophy, the process customers should also be involved from this phase at the latest in order to promote a benefit-oriented development. Process customers can be stakeholders such as controlling, purchasing or human resources departments, but especially the users of the later solution, thus also external customers or project partners, if applicable. If, for example, the situation in the company is that IT development work is basically outsourced to the software provider, then the software provider should be involved because it will have a significant and success-critical share in future IT projects. This example is representative of the value creation networks that are increasingly found in project work.[45]

In the (participatory) conception of the target state the future processes, the methods and tools used, including IT, the role model with its descriptions of tasks, responsibilities and authorisations, data used as well as the conception of the migration from the old to the new state (organisation, technology), all play an important role. Once the elaborated changes have been decided – here the client plays the central role – the organisational and technical requirements can be implemented as previously planned. The upcoming transition to the new procedures should be communicated to the relevant stakeholders at this point according to the communication concept, so that they can adjust to it.

With the *start of production* it is finally applied, initially e.g. in the selected pilot area. In accordance with the idea of early participation of users the transition should not represent an informational or know-how barrier. Nevertheless, as with any product launch, intensive support in the first weeks of application is recommended to avoid unnecessary errors, additional work and ultimately acceptance problems due to the novelty of the procedure. Last but not least, problems can be immediately identified, addressed and quickly remedied or at least integrated into the next project cycle.

Lean management lives from the orientation towards the benefit for the customer and the user. In addition to the usage, the value add that can be achieved with new methods etc. must therefore also be identified and ideally measured. A standardised key figure system for quantifying performance or its development, as exists in the production sector, is not known in PM. Nevertheless, the work in the PPM Laboratory at the University of Applied Sciences Central Hesse, a number of key performance indicators were identified that are considered useful by users (see e.g. Figure 7.12).

The respondents of a medium-sized production company considered the following key figures to be particularly relevant:[46]

---

45 cf. Wald/Schoper et al. 2015.
46 For definitions of the other key figures, see. Haemer 2019, pp. 86–105.

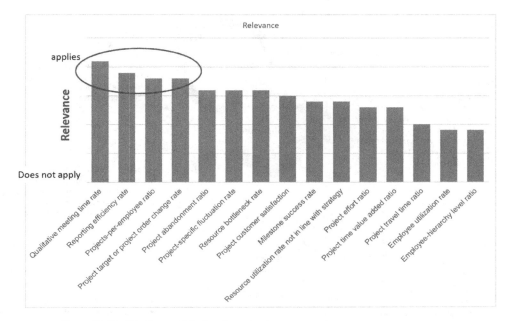

FIGURE 7.12   Lean PM indicators and their relevance – an example.

Reporting efficiency quota	Sum of actually used reports divided by the number of reports
Project per employee ratio	Arithmetic mean of the number of projects per employee
Project target or project order change rate	Number of project order changes divided by actual project duration
Qualitative meeting quota	Average share of relevant meeting time of the project members

Assessing whether indicators are suitable and useful involves more than just assessing their relevance. Typically, aspects of measurability, availability, influenceability and significance are also important. For example, a key figure that is relevant but for which no data can be measured cannot be used in practice. Here, therefore, it is necessary to define key figures that are practicable from the above-mentioned points of view and that are to be pursued. An important point of reference for this is the project mandate or the expected benefit of the measure articulated by the client in the early project phase (e.g. in the business case).

In terms of change management, the identified successes should be communicated to the organisation as early as possible during the application of the Lean PM concept. Potential ideas for further improvement, but also possibilities for expansion, should be incorporated into the backlog of the Lean PM implementation, so that the process described can be implemented in the sense of the PDCA cycle (Plan – Do – Check – Act). Figure 7.13 summarises the details of the process model.

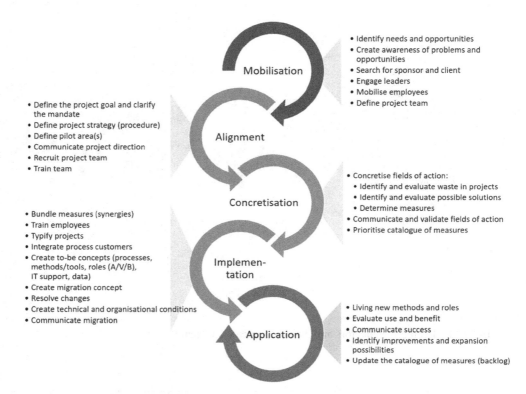

FIGURE 7.13 Process model for the introduction of Lean PM activities.

7.2.3 *Practices in project implementation*

When implementing Lean PM, some of the practices presented in this book can and should be used – true to the motto "Walk the talk!" – do what you talk about. Related to the practices presented in Figure 7.13 the following methods can be identified:

Mobilisation	SWOT analysis ... for the general identification of needs and opportunities
Alignment	Hoshin Kanri ... to align the organisation with the goals
	Lean PM maturity analysis ... for the technical concretisation of the fields of action
	Adaptation of the PM system ... to the appropriate selection of the project approach
	Agilometer ... for the concrete analysis of the control method
Concretisation	Value stream mapping (use of UPMF, Process Canvas/Makigami) ... to analyse the project and PM processes
	Gemba (Go on the spot) ... to understand and see the problems and possible solutions
	Root cause analysis (Ishikawa diagram, 5 Why question technique) ... for sustainable elimination of problems, incl. bundling of measures
	Prioritisation (TVD, WSJF) ... for the subsequent cost-benefit-oriented control of the introduction process, including the realisation of quick wins
	Breathing scope (MuSCoW rule, backlog management) ... to create scope flexibility in the management of the implementation process

Implementation	Process customer integration (participation) ... to ensure that the use and the utility of the solution are adequate
	Pareto principle ... to ensure an optimal cost-benefit ratio (overall view)
	Basically, all principles and methods of Lean PM ... to find solutions
Application	Key figures ... to measure the effects
	Standardisation ... to increase process reliability, quality and efficiency and as a defined starting point for further development
	Benefit revision ... to assess the effects and any gaps that may exist
	CIP (Kaizen, Starfish) ... for the continuous development of existing solutions (process, methods, tools, etc.)

This list does not claim to be exhaustive (as, in principle, the set of practices cannot be described exhaustively and fully) and can be adapted in the concrete project against the background of personal preferences, existing knowledge and general framework conditions of the organisation.

7.2.4 Success factors in the implementation of Lean PM

Some authors have empirically investigated the critical success factors and also the obstacles to the introduction of Lean PM.[47] The following factors from case studies, from field trials and from statistical analysis have emerged. The capabilities and opportunities for individuals and organisations to follow the roadmap will vary depending on the position and circumstances. But as far as possible, the following should be done to implement Lean projects:

- Select stakeholders (project managers and teams, departments or service providers, suppliers) who are willing and able to apply Lean Project Management practices. Ensure management support.
- Customer orientation both internally and externally: Structure the project organisation in such a way that downstream actors are involved in upstream processes and vice versa.
- Apply Target Value Design: Define and align project scope, budget and schedule to deliver value to customers and relevant stakeholders while challenging previous best practices.
- Ensure that resources (money, staff, schedule, etc.) can be moved across organisational boundaries to achieve the best returns at project level.
- Encourage cautious experimentation: Explore the adaptation and development of methods to pursue Lean PM. Use failure as an opportunity to learn rather than an opportunity to punish the guilty. Perceive mistakes as opportunities.
- Set-based concurrent engineering: Making design decisions at the Last Responsible Moment (LRM) – with explicit generation of alternatives and documented evaluation of these alternatives based on defined criteria.
- Practise project management according to lean principles: E.g. make the work process predictable, use pull systems to avoid overproduction and overload, if possible, analyse issues and problem reports at the point of occurrence, Poka-Yoke and 5S techniques, etc. Ensure basic understanding of the lean methods used.

---

47 See Ballard et al. 2007, p. 136 ff. or Busse 2017, p. 130 ff.

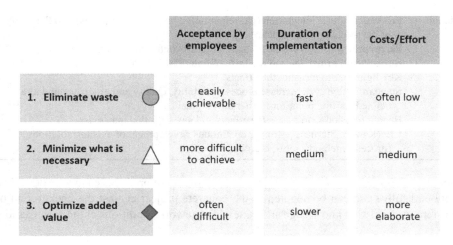

	Acceptance by employees	Duration of implementation	Costs/Effort
1. Eliminate waste	easily achievable	fast	often low
2. Minimize what is necessary	more difficult to achieve	medium	medium
3. Optimize added value	often difficult	slower	more elaborate

FIGURE 7.14   Basic strategy of process improvement.

Source: Based on Tie 2020, p. 7.

- Use of *first run studies*: Checking process capability in processes of technically progressive result development (conception, realisation) as well as PM in order to fulfil safety, quality, time and cost criteria. Standards as a recurring starting point.

In the first step of process improvements, the focus should always be on eliminating pure waste. The second priority is then to minimise business- or process-related waste and only then to optimise the value creation (see Figures 7.14).

Eliminating pure waste in the processes often succeeds faster and with less effort or costs, and acceptance for the change measures is easier to achieve due to the often existing obviousness. Secondary types of waste are characterised by the fact that they do not directly create customer benefits in the process, but are necessary due to higher-level requirements, e.g. company-wide controlling. Changes are therefore more difficult to achieve overall. Finally, the process steps that have already been identified as adding value need to be optimised in detail. With this described ranking, a potential-oriented approach can be taken.

In contrast, typical disruptive factors in implementation that need to be avoided can be identified:

- Culture changes too quickly
- Targeting cost reduction alone
- Poor employee participation
- Lack of resources
- Wrongly distributed rights and duties
- Incomplete competence formation
- Only intra-departmental communication
- Other internal barriers, e.g. demographic trends, job security, etc. which are not directly related to Lean PM but influence the project.

8

PERSPECTIVE OF LEAN PROJECT PORTFOLIO MANAGEMENT

> After reading this chapter, you will have a brief overview of ...
>
> - What project portfolio management (PPM) is all about,
> - typical criticisms of traditional PPM and empirical success factors, and
> - initial approaches to agilising PPM using core lean principles.

8.1 Background of project portfolio management

The management of entire project landscapes in organisations is referred to as *project portfolio management* (PPM). Especially in large companies, PPM is a proven tool for the management level to steer the organisation according to the strategy.[1] The average project portfolio in a comprehensive multi-project management study by the Universities of Technology in Darmstadt and Berlin...

- has 50 projects,
- has an annual budget of 30.5 million euros,
- consists of 40% must-do projects,
- 65% of the budget is used up by old projects,
- includes 10% absolutely novel and 38% routine projects.[2]

---

1 see Wagner 2016.
2 see Gemünden et al. 2013. In the study 189 project portfolios with over 28,000 projects and a project volume of 21.885 billion euros were considered.

DOI: 10.4324/9781003435402-9

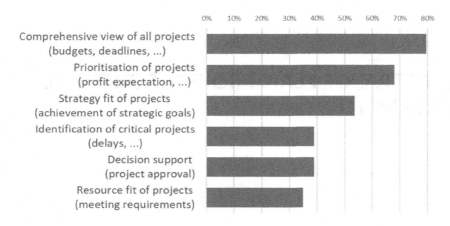

FIGURE 8.1 Why project portfolio management (PPM)?

A core task of PPM is to design a project portfolio that optimally serves the strategic orientation of the company – under the resources used. A PPM study by Scheer shows further current objectives of PPM (see Figure 8.1).[3]

The implementation of PPM tasks currently appears to be much more difficult than it was a few years (decades) ago. Increasing dynamics and complexity (e.g. digitalisation) characterise the framework conditions of modern corporate management, which ensure that the demands on PPM have become more unstable.

PPM is a young management discipline. The first German standard for the domain in the form of *DIN 69909* was published in 2013. Accordingly, the fundamental need for further development of this domain is naturally high, especially since the existing standards do not yet reflect the increasing trend of agilisation. These national and international standards include *ISO 21504*, which was adopted by DIN in 2017, supplemented by *ISO 21505* on the topic of governance, the *Standard for Portfolio Management of* PMI which is published as the US ANSI standard, *PMI FS-POM-2006*, the *Individual Competence Baseline for Portfolio Management of* IPMA from 2015, the *Management of Portfolios* (MoP) by AXELOS Limited – *P3O* (*Portfolio, Programme and Project Offices*), published in 2008 as the first version by the British Office of Government.[4]

As a result, there is no uniform international standard for PPM. The works mentioned have different focal points on PPM – from the perspective of corporate strategy development, roles, processes and methods to the technocratic definition of terms. As a common denominator, as it were, a high-level process model can be synthesised (Figure 8.2), which, supplemented by the PPM support processes, provides the starting point and the regulatory framework for the adaptation of Lean Management principles and practices in the sense of a modern PPM.

---

3 see Scheer 2017, p. 8.
4 cf. Klotz/Marx 2018, pp. 26–35; Süß 2016, pp. 126–132.

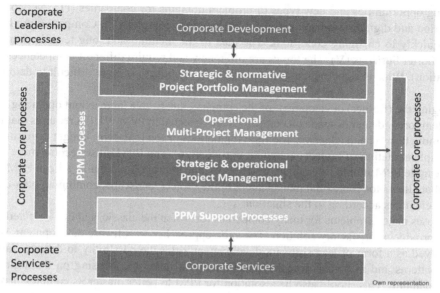

FIGURE 8.2 High-level process model PPM.

The PPM process model, based on the St. Gallen Management Model (see Section 7.1.2), shows the basic division of PPM into strategic, normative and operative management processes, supplemented by the technical processes of individual project management and the enabling support processes. The processes described in Figure 8.2 in the current version (2023) include ...

Strategic & normative PPM	PPM System Strategy Determination, Project Portfolio Authorisation, PPM Governance
Operational PPM	Project Demand Management, Resource Management, Performance Management, Benefits Management
PPM Support	Development of PPM Methods & Tools, PPM System Operations

The above-mentioned levels are supplemented by cross-sectional processes of Information & Knowledge Management, Change and Risk Management as well as Requirements Management, which is located at the level of individual measures. The PPM processes are integrated into strategic corporate development, the business areas and the general corporate services.

Volatility, uncertainty, complexity and Ambiguity (VUCA) describe the difficult framework conditions of modern corporate management in a rapidly changing business world.[5] The digital transformation of companies in particular serves as an example of this, which, for example in the area of the so-called *Fourth Industrial Revolution* (4IR, or Industry 4.0), leads to rapidly

---

5 see t2informatik n.d.

developing opportunities and new types of problem domains for companies (IT). Networking, globalisation and digitalisation are making solutions more complex (C). Companies are forced to react quickly to changing conditions. The idea of the medium and long term is changing, towards shorter horizons (V). The topics are often new and require different competences (e.g. digitalisation). What is right/meaningful or wrong (in the future) is more difficult to determine (U, A).

Managing the (further) development of the company means to a large extent operating PPM. The contemporary demands made on companies are reflected in PPM. VUCA requires that PPM itself be more flexible and adaptive. Too rigid regulations or too long-term budget commitments for the implementation of projects are counterproductive.[6] The PPM must become more agile, or a more agile PPM should be enabled, if necessary and appropriate. Finally, a Lean PPM must emerge that allows for contextual management of the project landscape, combining plan-driven and agile elements according to the situation.

This poses major problems for many companies for which the classic, stability-oriented PPM does not provide a satisfactory solution. The agile approaches to software development, which have evolved as a form of lean software development, offer the possibility to discard classical thought patterns and to better meet the requirements of today by focusing on flexibility and iterative planning. This should also be a solution for PPM to better master its complex tasks. In this context, findings from individual project management show that the models are often applied in a mixed form, as this often corresponds better to the framework conditions in the companies. Therefore, hybrid project approaches – such as Lean PM – have developed that combine the plan-driven approach with agile methods.[7] This ultimately creates multi-modal project landscapes that need to be managed. The mixing of project approaches with different planning and control philosophies in one portfolio results in new challenges for the design of PPM.[8] While classical methods focus on milestones and *a priori* defined deliverables, agile approaches postulate a fundamental openness to results. It is therefore necessary to manage both approaches in the overarching control of the PPM.

The aim is therefore to develop and shape a conceptual approach for a modern PPM using Lean Management principles and practices. The central question in designing a lean PPM approach is: *How can Lean Management principles and practices be applied to the processes, structures and methods of project portfolio management in order to improve its efficiency and effectiveness – not least in the context of the VUCA-requirements?*

The lean elements should be systematically mapped to the sub-domains of PPM such as project proposal management, project prioritisation, portfolio balancing, etc.,[9] and to search for meaningful and beneficial applications. The result is recommendations for action at the strategic, tactical and operational levels of PPM.

---

6 see Schnichels-Fahrbach/Munz 2016.
7 see Blust 2019.
8 see Wagner 2016.
9 see Lock/Wagner 2019.

8.2   Criticism of the classic PPM

The classic PPM can be characterised as stability-oriented PPM. It usually starts with a project proposal that is already developed at an early stage to such an extent that a quantitative-qualitative utility analysis can be carried out on its basis. This often entails a number of problems. For example, the integration of the application process into the usually annual budgeting process requires very early processing. Here is an example of a process.

PROCESS EXAMPLE

- In order to have the budgeting for the business year 20YY+1 approved with effect from January of the year, the project applications for the year 20YY+1 must already be submitted, or at least initiated, in August. It must therefore be clear by the middle of 20YY which projects are to be implemented in 20YY+1. The application must be complete by the end of Q3/20YY at the latest.
- Experience shows that the senior management, which makes the portfolio decisions, often demands concrete and detailed decision documents (= project proposal) from the applicants.
- The preparation of a complete project proposal also brings with it the problem that key figures relevant to decision-making, especially the budget, have to be stated before the project and the associated increase in knowledge. It is therefore pre-programmed that (in retrospect) wrong figures are used as a basis for decision-making – and basically the management accepts this with their eyes open (waste).
- The correction of the key project parameters (costs, deadlines), which is often due, requires permanent readjustment during the course of the financial year/project execution, including evaluation of the business case. This may lead to a completely new situation, which may even require the project to be cancelled, which is an immense waste. Even – and especially – if a project is not cancelled, despite knowing better, this leads to even more wasted effort (sunk cost issue). Some major public and private sector projects in Germany (airport BER, SAP implementation at Lidl, Haribo, Deutsche Post, etc.) are striking examples of this.[10]

In classical PPM organisations – not least the applicants – tend to make their project proposals as big as possible, i.e. (supposedly) important. This causes the following problems:

- The complexity of the project increases and the PM and the technical work becomes more difficult.[11]
- The risk of failure or at least less success increases – large projects fail more often.[12]

---

10 see e.g. Kroker 2018; Brandenburg et al. 2020.
11 see Boehm/Turner 2009.
12 see Standish Group 2015.

- More staff are tied up by the project,

  - the projects take longer
  - the teams are bigger
  - the coordination effort becomes (exponentially) greater
  - ... and are not available for other tasks or harmful multitasking is the result.

- Due to the larger scope, the probability of change requests and thus the effort for the scope/ CR management increases.
- The psychological, possibly political hurdle for a justified project break-off becomes greater, since a lot of money has already flowed into the project (sunk cost bias).

8.3   Success factors in project portfolio management

What makes top performers in multi-project management? The study by Gemünden and Kock regularly examines success factors in the design of PPM in companies.[13] Here are the findings from the 2016 study with 134 project portfolios examined: Top performers (of PPM) ...

- use visualisations to support decision-making ($\rightarrow$ Visual Management),
- make timely portfolio decisions ($\rightarrow$ flow),
- communicate the result of important portfolio decisions uniformly and across all hierarchical levels ($\rightarrow$ Hoshin Kanri),[14]
- formulate clear goals that are well communicated ($\rightarrow$ Hoshin Kanri),
- live a clearly defined multi-project management process ($\rightarrow$ standardisation),
- set higher standards for individual PM ($\rightarrow$ standardisation),
- use a formal risk management process ($\rightarrow$ standardisation),
- actively involve the stakeholders in the strategy process ($\rightarrow$ customer orientation),
- do not evaluate errors as failure ($\rightarrow$ error culture, Hansei),[15]
- consciously withhold resources for more flexibility,
- monitor their portfolio more frequently and systematically ($\rightarrow$ flow, pulsing),
- react significantly faster to changed conditions ($\rightarrow$ flow, First Defect Stop),[16]
- give project managers the opportunity to play an active role in shaping the portfolio ($\rightarrow$ Kaizen),
- require well-defined and validated business cases ($\rightarrow$ benefit orientation),
- make decisions on a rational basis using facts, data and figures ($\rightarrow$ use of key figures)
- use predefined budget buckets (Strategic Buckets) to maintain a balanced portfolio. Especially in times of high market turbulence, the use of predefined budget buckets becomes more and more important.

---

13 see Gemünden et al. 2016, pp. 20–96.
14 see also Kammerer et al. 2012, p. 433.
15 see Bertagnolli 2018, p. 354 f.
16 cf. Gorecki/Pautsch 2013, p. 57.

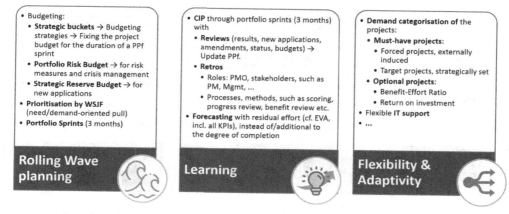

- Budgeting:
  - **Strategic buckets** → Budgeting strategies → Fixing the project budget for the duration of a PPf sprint
  - **Portfolio Risk Budget** → for risk measures and crisis management
  - **Strategic Reserve Budget** → for new applications
- **Prioritisation by WSJF** (need/demand-oriented pull)
- **Portfolio Sprints** (3 months)

Rolling Wave planning

- **CIP** through portfolio sprints (3 months) with
  - **Reviews** (results, new applications, amendments, status, budgets) → Update PPf.
  - **Retros**
    - Roles: PMO, stakeholders, such as PM, Mgmt, ...
    - Processes, methods, such as scoring, progress review, benefit review etc.
  - **Forecasting** with residual effort (cf. EVA, incl. all KPIs), instead of/additional to the degree of completion

Learning

- **Demand categorisation of** the projects:
  - **Must-have projects:**
    - Forced projects, externally induced
    - Target projects, strategically set
  - **Optional projects:**
    - Benefit-Effort Ratio
    - Return on investment
- **Flexible IT support**
- ...

Flexibility & Adaptivity

FIGURE 8.3   Impulses for operationalising agile design principles in PPM, part 1.

This list represents the selection of insights that can ultimately be associated with practices that are also used in Lean Management, as already presented in the enumeration. This particularly motivates the development of the *Lean-Agile PPM* approach, which systematically adapts the principles and practices of Lean Management to PPM.

A direct, explicit design of Lean Project Portfolio Management (Lean PPM) is not yet known.[17] However, in the course of the agilisation of PM and the scaling of projects, there are concepts that extend into PPM in terms of their technical and organisational scope. First and foremost is the SAFe concept, which itself also speaks of Lean Portfolio Management – without systematically deriving this. Important elements of SAFe at the portfolio level are value-added orientation, flexible budgeting of projects, regular delivery of results, pull-oriented processing of measures (IT-Kanban) etc.[18]

An important component of Lean PPM should be the strengthening of flexibility and adaptivity – i.e. agilisation – of the PPM (if this makes sense). Figure 3.2 already shows the benefits that can be derived from agilisation of the "PPM system": *Robustness, effectiveness, innovation-friendliness* and *responsiveness* lead to improved goal achievement in the dynamic, complex PPM. A purely agile approach stands, so to speak, for the opposite pole of a classic, plan-driven hierarchical system, which presupposes relatively stable states. As in Lean PM, Lean-Agile PPM combines both poles in the sense of a hybrid concept that operates under the guiding principle of Lean Thinking.

8.4   Approaches to increase the agility of project portfolio management

Lean-Agile PPM means, last but not least, increasing the adaptivity and flexibility of PPM compared to the classical stability-oriented, i.e. rather static PPM. It is therefore advisable to

---

17 but is work-in-progress at the PPM Laboratory of the University of Applied Sciences Central Hesse (GER).
18 see Mathis 2018; Komus et al. 2020; Leffingwell 2018; SAFe 2020.

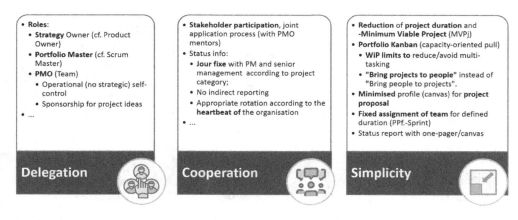

FIGURE 8.4 Impulses for operationalising agile design principles in PPM, part 2.

look for possibilities to agilise PPM. The derived design principles of agility (see Figure 3.2) – cooperation, delegation, learning, rolling wave planning and simplicity – can be used to identify concrete approaches (see Figure 8.3 and Figure 8.4).

The SAFe concept also identifies a number of operational elements for agilising PPM: Artefacts (Backlog, Portfolio Kanban, Portfolio Vision), practices (KPIs, Epics and Enablers, Value Streams, Lean Budgets) and roles (Epic Owner, Enterprise Architect). The practices shown in Figure 8.3 and Figure 8.4 are intended at this point as suggestions for the reader. A complete discussion is beyond the scope of this book – however, individual elements have already been presented in the context of Lean PM or will be taken up in the following.

8.5 Application of the lean core principles

8.5.1 Basics

In the first step, the values of Lean Management must be fundamentally transferred to PPM (cf. Lean PM basics). What do the core terms mean in the context of PPM?

Waste in PPM?
The types of waste in Lean Production are not to be used one-to-one. Rather, those types of waste that can generally be identified in the area of office floor processes (vs. shop floor) apply in the *secondary areas* (cf. Lean Administration, Lean Office).

As in Lean PM, this also includes the areas of documentation & data processing, processes & organisation, communication, planning & design and service provision. In the technical sense, the latter takes place in the projects themselves – e.g. the development of a product – but a primary value stream can also be identified in PPM, namely the end-to-end value creation process "from the project idea to the realisation of benefits".

Customer in PPM?

The concept of the customer in the field of PPM should also be understood in a broader perspective – and not only include the company's customers. As in Lean PM, it makes sense to look at the stakeholders of PPM. By analogy, those stakeholders should also be regarded as customers here who ...

- (directly or indirectly) obtain a benefit from the PPM or
- have commissioned them (internally or externally) or
- (formally or informally) have a high influence on the design of the project portfolio and its management system (PPM).

Generally speaking, the customers of PPM are the process customers of any business process of PPM.

Value streams in PPM?

The identification of the value streams of the domain under consideration for the analysis and elimination of waste is a core task of Lean Management. In this respect, it is also very important in Lean-Agile PPM.

The value streams, i.e. the value-added processes of the PPM can be identified and processed with the help of the PPM process model. The primary value creation process of PPM has already been identified with the end-to-end process "From project idea to benefit realisation". This is the most meaningful process of PPM, because it encompasses the procedure from the identification of a business need, through the processing of a corresponding solution and its commissioning (or availability), to the determination of the realisation of benefits. Other value streams are identified in the following section.

Flow principle in PPM?

In PPM, the consideration of flows refers to information flows. Physical flows, such as in the creation of a physical product (building, plant, etc.) take place in the projects themselves and are therefore not the subject of PPM at this point. The flow principle here generally demands minimum throughput times, low waiting times in a transparent flow of information – without interruptions, misinformation or backflows. Thus, the requirements apply as in every administrative and controlling process of corporate management that is designed according to lean principles.

Pull principle in PPM?

The pull principle describes the flow of a value stream in which an upstream unit only produces if the units immediately downstream announce a demand. As can already be seen in the area of Lean PM, this demand-driven pull principle is not always sensible to implement one-to-one in the area of management processes. Nevertheless, useful areas of application can be identified within PPM.

In particular, the capacity-oriented pull principle must be applied here, because, by analogy, only as many tasks as the work-in-progress limit allows are processed simultaneously by the processing unit. The unit, e.g. an organisational unit, pulls new tasks from the backlog only if it has the capacity to do so – in the PPM context, projects.

Perfection in the PPM?
In the context of PPM, striving for perfection can only mean avoiding waste of any kind in business processes, as completely as possible, and thus achieving the maximum possible efficiency. In PPM, this includes achieving an optimal cost-benefit ratio in the design of the project portfolio, i.e. in the selection of projects. In general, it should be noted that the benefit of a project cannot always and only be measured in terms of monetary value added, since qualitative and strategic aspects are also relevant in modern PPM.[19]

Interpretation of the core principles of PPM
In the second step, the described core principles of Lean Management are interpreted in concrete terms with reference to PPM. In general, the system followed is:

1. Who are the customers of the PPM or its elements?
2. What benefits do they expect from the PPM?
3. Which value creation processes (value streams) generate the services required for this?
4. What types of waste can be identified in the processes in general?
5. How can the flow principle be implemented in the value stream processes?
6. How can the pull principle be implemented in the value stream processes?
7. What does perfection mean in the context of performance in PPM?

The process model for multi-project management developed in the PPM Laboratory of the University of Applied Sciences Central Hesse serves as the regulatory framework (see Section 8.1). This makes it possible to not start with the conceptual considerations of a greenfield site. Ultimately, it must be evaluated that all processes of the model contribute (at least indirectly) to value creation and that, on the other hand, no further processes are necessary for value creation, i.e. that they are missing from the model.

8.5.2 Customer and value in PPM

The identification of the customers of the PPM is done on the basis of the PPM processes. This means that a broader understanding of the term is applied than focusing solely on (real) end or corporate customers. The customer concept thus refers to the process customers (see Section 8.5.1). Figure 8.5 lists the typical customers and those responsible for implementing the PPM processes.

---

19 cf. Schütte et al. 2019.

		Shareholder	Senior Management	PPF-/CPMSys Manager	Division Manager	Project Manager	Resource Manager	Knowledge manager	Change Manager	Product Manager	Staff/Team
Strategic & normative PPM:	PPM System Strategy Determination		S	R							
	Project Portfolio Authorisation	S	S	R							
	PPM Governance		S	R							
Operational MPM:	Project Demand Management			R	S						
	Performance Management		S	R	S	S	S				
	Resource Management			R		S	S				
	Benefits Management		S	R	S			S			
Project Management:	Requirements Management				S					R	
	Project/phase initialisation			S	S	R					
	Project/phase planning			S		R					
	Project/phase control			S		R					
	Project/phase completion			S	S	R				S	
PPM Support:	Development of PPM Methods & Tools			R		S					S
	PPM System Operations			R		S					S
Overarching	Information Management					S			R		S
	Risk Management										
	Change Management				S				R		S

FIGURE 8.5   RS matrix for PPM.

The *RS matrix* is a proposal to extend or focus the well-known RACI matrix by the identification and documentation of the process customer relationship ($\rightarrow$ *SICAR*).[20] Figure 8.5 shows the focal points with regard to the customer relationship ("S" for served/ satisfied) and the implementation or process responsibility ("R") with a view to the typical relevant roles. In detail, of course, there may be deviations, e.g. the separation of PPM and an operational Project Management Office or the combination of divisional and resource responsibility. In this sense, the RS matrix shown can serve as a blueprint for company-specific design.

With the identification of the customers of the PPM processes, their value concepts can also be derived and concretised. In general, the following overall corporate value definition in Lean PPM results:

- In the PPM, the value consists of selecting and managing projects that best serve the overall business objective.
- The aim of PPM is to ensure that selected projects, as opposed to other projects, generate the maximum strategic benefit.
- Value is created in the individual project context through the fulfilment of the project mission.

The reasons for PPM listed in Figure 8.1 are to be classified as a means to the end of this overarching perspective. After the following identification of the value streams in PPM, we come back to the associated customer benefits.

20 Remark: To avoid unfortunate associations, we have reversed the acronym.

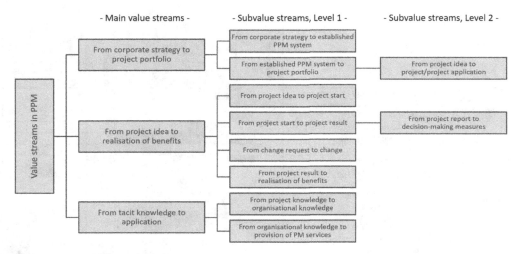

FIGURE 8.6   Value streams in PPM.

8.5.3   *Value streams in PPM*

The flow principle demands a process-oriented design of value creation. The optimisation of the identified value streams is a central principle of the implementation of Lean Management. They are to be identified with the help of the high-level PPM process model and designed with its sub-processes. The following main and secondary value streams can be identified (see Figure 8.6).

In order to analyse the identified PPM value streams and to be able to optimise them in the sense of Lean Management, the associated customer benefits must first be identified. Value streams that can be assigned to individual project management will not be discussed further at this point (Section 4.9 is referred to).

From corporate strategy to project portfolio

The activated project portfolio offers optimal implementation of the agreed corporate strategy with the resources deployed

- Sub: *From corporate strategy to established PPM system*
  The established PPM system is structurally suitable for organising the achievement of the overarching goal of an optimal portfolio.
- Sub: *From the established PPM system to the project portfolio*
  The activated project portfolio is suitable for achieving the overall objective of implementing the corporate strategy.
- Sub: *From the project idea to the proposed project/project application*
  The decision to implement an idea is made promptly; resources are provided appropriately.

From the project idea to the realisation of benefits
The desired result of a project is available in a timely and relevant manner, is used and delivers the expected added value.

- Sub: *From project report to decision-making measures*
  Escalations from within the project are resolved in an appropriate and timely manner, usually as quickly as possible – taking into account the overarching priorities.
- Sub: *From Change Request to Change*
  Change Requests related to the project order (outside the agreed tolerances) are decided on in an appropriate and timely manner, usually as quickly as possible, and authorised in a positive case.
- Sub: *From project result to benefit realisation*
  The benefits associated with the project are optimally realised through the use of the project results.

From tacit knowledge to application
The technical and methodological knowledge gained in the project work is made optimally applicable for further use in the organisations's (IT) services.

- Sub: *From project knowledge to organisational knowledge*
  The technical and methodological knowledge gained in the course of the project work is made available in an optimal way for further use across the organisation.
- Sub: *From organisational knowledge to the provision of PM (IT) services*
  The technical and methodological knowledge gained in the project work is made optimally applicable for further use (IT-supported).

With the help of the PM process model, the value streams can be designed. For example, the value stream "From the project idea to the project start" is roughly orchestrated as follows:
$\bigcirc \rightarrow$ *[Idea Management]* $\rightarrow$ *[Project Initialisation]* $\rightarrow$ *[Multi-PM Evaluation]* $\rightarrow$ *[Resource Management]* $\rightarrow$ *[Portfolio Alignment]* $\rightarrow$ *[Project Planning]* $\rightarrow \bigcirc$

8.5.4 Waste in PPM

Avoiding waste is the central paradigm of Lean Management. In the field of PPM there are a number of specific wastes to be identified – on top of the usual typical wastes in administrative processes. These include:[21]

- Bureaucracy, e.g. in the form of complicated project applications ... leads to extra work and unacceptance.
- Project decisions based on faulty assumptions (e.g. costs too low) ... lead to undesirable developments.

---

21 also inspired by multi-project management study Gemünden et al. 2016.

- Delays due to untimely project idea development and implementation ... lead to reduced business value.
- Frequent corrections of project orders (change requests) ... lead to additional work and irritations.
- Projects that are not cancelled (sunk costs) ... lead to a waste of money and resources and thus additionally to opportunity costs due to lost alternatives.
- Harmful multitasking ... causes friction losses (set-up times).
- Late realisation of benefits through long-running projects ... leads to deterioration of the business case.
- Unnecessary complexity of projects and the project landscape ... leads to more difficult coordination.
- Inadequate evaluation standard (scoring model) for projects ... leads to incorrect prioritisations.
- Too complicated scoring model ... leads to intransparency and thus unacceptance of project prioritisation.
- (Unbalanced) separation of PPM and business responsibility ... leads to unacceptance and "submarine projects" (bypassing the process).
- Lack of prioritisation ... leads to misallocation of resources.
- Incorrect project status reports ... lead to mismanagement.
- (Unbalanced) separation of project implementation responsibility and PPM ... leads to loss of information.
- Lack of/insufficient senior management commitment ... leads to the undermining of the entire PPM system and thus total waste within the PPM.
- Lack of/insufficient knowledge management ... leads to inefficiencies and errors in the processing of upcoming project tasks.
- Lack of/insufficient change management ... leads to reduced application of project results and thus to reduced benefits of the same of the same.
- Insufficient IT support ... leads to poor information quality, time-consuming data preparation, delayed information flow.
- Missing/insufficient strategic integration of PPM ... leads to misprioritisation of projects and thus misallocation of resources.
- (Too) rigid and early project budgeting ... leads to inflexibility or additional work (change requests).
- (Too) long project durations ... lead to delayed, if not destroyed, realisation of benefits.
- Lack of/insufficient resource management ... leads to miscoverage of needs.
- Full utilisation of resources ... leads to inflexibility in the event of changes in demand.
- Inefficient PPM meetings ... lead at least to additional work.
- Lack of/insufficient risk management (in the sense of hazard management) ... leads to increased and expensive crisis management.
- Lack of/inadequate opportunity management ... leads to non-realisation of business ideas, investment opportunities or strategy changes.
- Too long cycles of project portfolio review ... lead to non-recognition of undesirable developments.
- Faulty/insufficient Business Case management ... leads to non-recognition of undesirable developments.

- Missing/insufficient knowledge preparation posthumously ... leads to wasted learning effects and thus to waste in follow-up activities.
- etc.

8.5.5 Flow principle in PPM

Ideally, a process should run at the same pace as on a conveyor belt – a stream that flows at a constant speed until the result finally reaches the process customer. The flow does not back up at any point ("There are no dams.") but is also not accelerated strongly at any point ("There is no waterfall."). Waterfalls and dams cost additional money, are a waste and should therefore be eliminated.[22] The flow objects considered in the process are generally material and information – in the area of management processes (such as PPM) only information.

Two specific principles of action that enable the implementation of the flow principle are – in addition to the pull principle – the *One-Piece Flow* and the *First Defect Stop*. In the area of administrative and management processes, the Makigami diagram is used as a method of analysis (see Section 4.5). The previously identified value streams are therefore suitable for such an analysis. The use of the Makigami diagram will not be discussed further here, as it is a generic tool that can be used in PPM as well as in any other business process domain. What is more interesting is the question of whether and how *One-Piece Flow* and *First Defect Stop* can be applied as principles to PPM in a meaningful and beneficial way.

8.5.5.1 First Defect Stop

When using the First Defect Stop, a process step is provided with an automatic stop as soon as an error is detected. The responsible employee then has the task of analysing the error and rectifying it within the scope of the defined responsibilities – starting with himself.[23] In the field of PPM this can be applied to faulty information that prevents subsequent steps in the process from being carried out. Important information flows in PPM are the project application process and reporting. Applicants are often not practised in processing project applications, so they may provide incorrect or incomplete information. Instead of continuing to work with erroneous information, the parties involved in this process – the applicant and the project portfolio manager as the recipient of the information – should clean it up immediately to avoid unnecessary complications in the follow-up process (the approval). If applicable, the Project Management Office should closely supervise the project applicants in the business departments – and not allow a "throwing over the fence" of supposedly finished project applications.

Another application of the First Defect Stop can be identified in dealing with projects that have become unbalanced. This imbalance usually manifests itself in corresponding status reports in which individual positions are set to red. Such situations should not be accepted and merely treated with stronger monitoring. Instead, an immediate solution – for example, with the help of a task force – is advisable (see Section 5.1.5). This can be defined as a supporting measure within the project, but also by the PPM. In the case of serious complications, one should not be afraid to

---

22 after Gorecki/Pautsch 2013, p. 29.
23 cf. Gorecki/Pautsch 2013, p. 57.

put a project on hold in order to give time and priority to solving the problem. Otherwise, as in the production process, further complications can arise in the subsequent course of the project, which can be more serious overall and cause delays/empty runs, duplicate work, etc.

8.5.5.2 One-Piece Flow

In the One-Piece Flow, the process carried out resembles a human chain that passes the water buckets one by one from person to person to (quickly) extinguish a fire. In this way, water arrives quickly at the destination, an initial benefit is achieved quickly and, if necessary, the chain of people can also be adjusted locally as required. In particular, waiting times (= waste!) of subsequent (process) chain links are reduced. However, this batch size 1 is not always worthwhile – if set-up times speak against it, the optimum lies in a compromise.

In classical PPM it is possible to identify places where the opposite of One-Piece Flow, i.e. block processing, is usually practised. In block processing, processing is only started when a certain amount (> 1) of input is available or a correspondingly defined point in time is reached. In PPM, the following can be identified here:

- Working on large projects (with many individual topics)
- The processing of all project applications (approval) at a certain point in time (e.g. once a year).
- The processing of all status reports to be made according to a defined schedule (e.g. at the end of the month).
- The release of change requests through fixed portfolio board meetings (e.g. once a quarter)

Certainly, other similar situations can be found in companies. What they all have in common is that the feedback is delayed and not necessarily in line with the needs in order to be able to continue working steadily. A second, perhaps subordinate, but nevertheless well-known effect could be the information and work overflow that occurs at certain deadlines, which tends to make the processing quality suffer.[24] But what could a One-Piece Flow look like in the contexts described?

1. The projects should be made smaller (narrower and shorter).

Project managers often tend to make their project large and thus supposedly important ("A-projects"). Small projects, however, are quicker to navigate through the organisation, tie up fewer resources at all times, deliver faster results and can be better readjusted and possibly also stopped. Of course, you can't make a molehill out of a mountain and you can't scale down complex projects at will (see Section 2.3). But at least project sections can be defined, each of which delivers a self-sufficient result (e.g. a pilot application or a release). A proven rule of thumb is that a project or project section should not last longer than 9 to 12 months, ideally even only one quarter.

---

24 It is known from work psychology that the umpteenth decision a manager has to make in a day is often not made with the same decision-making ability as the first one of the day ("Barack Obama only ever wears blue suits."), see Eube 2016.

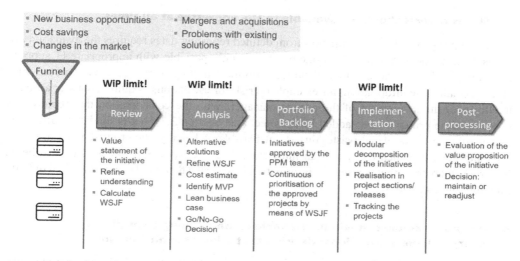

FIGURE 8.7 Project Kanban for the value stream "From project idea to utilisation".

The realisation of making projects as small as possible (One-Piece Flow) ultimately necessitates a redesign of the project categories A/B/C. A-projects are important, but not necessarily large (at least the individual project sections are not).

Project applications should be able to be submitted and processed without a fixed (one-off) deadline.

This means that a decoupling of the strict binding of the approval of projects with the budgeting of the financial year planning must take place. This can be achieved if so-called *Strategic Buckets* for the projects, divided according to their project subject, are formed within the framework of the annual budgeting (see Section 8.4). Such a "bucket" could be, for example, the one for the company's digitisation projects. It is easy to see that this is a very good way for senior management to specify and implement strategic preferences. An application of this idea is also provided by the SAFe concept with the *value streams* that are budgeted there.

The second well-known practice in this context is the use of project Kanban boards (see Section 4.5). These can be used to operationalise a continuous, funnel-like process (see Figure 8.7).[25]

The structure depicted in the project Kanban corresponds to the identified value creation process. The project Kanban board allows an inflow of project proposals at any time. Processing is then carried out according to the WiP limits by means of a capacity-oriented pull principle. The WiP limits refer, for example, to the personnel capacities available in the PPM. Urgencies can be taken into account through corresponding priority specifications. Together with the Strategic Buckets, a heuristic for implementing the flow principle is created.

---

25 in accordance with SAFe 2020 or Liechti 2019.

2. Status reports should be available to the addressee at all times.

The decoupling of status determination from defined reporting dates requires the availability of status information at any time. In principle, this is only possible with appropriate IT support, for example through access to accounting information (expenses, costs, etc.) of an ERP system. Qualitative information, for example on risks or escalations, requires (without artificial intelligence) the assessment of the project manager. With the consideration that reporting is not part of the primary value added by a project (see Section 3.3), but is fundamentally to be classified as process-related waste, this is not feasible. A healthy compromise must therefore be found here, in which the use of an integrated, database-based PM system will also be an important enabler, so that e.g. logged information is recorded there and can be retrieved directly (ubiquity of information).

3. Decision instances should be available on an ongoing basis, if necessary to issue a provisional decision that allows for further work.

It is the responsibility of the senior management to make the basic decisions regarding the project portfolio. However, the senior management of a company is usually not available all the time and portfolio board meetings with the protagonists are often planned on a fixed long-term basis following a certain cycle. In companies, one can encounter monthly, bi-monthly or quarterly cycles, depending on the availability, prioritisation and project orientation of the company or its management.

In the sense of maintaining the flow of decisions, this is partly too little: If a need for a decision arises on day 1 after a regular meeting, as a rule, one should not have to wait almost three months for it. Competencies must therefore be created here that enable prompt, immediate follow-up. The role of the project commissioner or strategy owner is predestined here. He or she should be available in a timely manner and be able to make fundamental decisions. If necessary, deputy and provisional arrangements can also suffice here, which enable work to continue and accept the possible damage in the event of a redisposition (cf. Small Steering Committee in Chapter 5).

8.5.6 Pull principle in PPM

The demand-driven pull principle in PPM can be identified in two areas: the budgeting process for the company in the following year, which entails the requirement to design the upcoming project landscape, and the scheduling of the productive availability of project results, which is internally desired (e.g. go-to-market time) but can also be externally induced (e.g. implementation of legal requirements).

With regard to the budgeting process, some considerations have already been made in the explanations on the flow principle. The hypothesis is that the budgeting process will continue to go hand in hand with the financial year planning in companies and thus likely take place once a year.[26] This process will therefore continue to "pull" on the availability of information

---

26 cf. Lingnau et al. 2004.

on required budgets. With a live, filled project Kanban board, the cross sum of the projects/ project applications in progress and in the pipeline can be drawn on each due date. Together with the formation of domain-specific Strategic Buckets induced by strategic considerations, project budgets including risk reserves for unforeseeable events (changes in ongoing projects, new projects during the year, etc.) can be defined at this level.

Last but not least, this can contribute decisively to resolving a phenomenon observed in practice, the *Knowing-Doing-Gap* of PPM. Everyone basically knows that the estimates of costs in project proposals are made to the best of one's knowledge and belief. Nevertheless, senior management regularly makes its individual decisions on the basis of these figures – with the almost inevitable consequences of readjustments, etc.[27]

A further application of the demand-oriented pull principle can be found in the scheduling of projects in the so-called *Project Roadmap*. The Project Roadmap is a high-level project schedule that often describes the temporal placement of the projects in a portfolio in the form of rough Gantt charts. A demand- or benefit-oriented move arises here through the varying urgency of the project results, as is also generally the case through the *Weighted-Shortest-Job-First* system (see Section 4.3). At this point I would like to adopt the system of Shenhar and Dvir, who classify projects as follows:[28]

- *Regular projects:*
  The completion date is not critical for success.
- *Fast, competitive projects:*
  Completing the project as soon as possible is important for the company's competitive advantage or market position.
- *Time-critical projects:*
  Meeting a deadline is critical for the project's success, missing it will cause the project to fail.
- *Flash projects:*

Projects in crisis situations, e.g. natural disasters or sudden pandemics.
The regular projects, e.g. projects within the framework of political education initiatives of *non-governmental organisations* (NGOs), do not exert any pull. The competitive projects, which are usually motivated by a desired *Return on Investment* (ROI), should be realised as early as possible in the portfolio, because the benefit generally accrues with the commissioning of the result, e.g. the market launch of a newly developed product. Time-critical projects require backward scheduling in relation to the desired or required availability date. Here, in general, the later to start, the better – as the probability of unknown developments at project maturity can be reduced, and especially since late cost emergence is generally preferable. An example of such a project is the implementation of legal requirements on a cut-off date, such as the double taxation agreement of Germany with Luxembourg on 01.01.2014.[29] Finally, lightning projects trigger an

27 cf. Hope/Fraser 2003.
28 see Shenhar/Dvir 2007, p. 127.
29 see BMF 2019.

immediate start, e.g. the rescue operation of children from the Tham Luang cave (Vietnam).[30] To sum up:

- ROI projects as early as possible!
- Appointment projects as late as possible!
- Start lightning projects immediately!

Furthermore, a capacity-driven pull can be usefully applied to avoid overload due to multitasking. The project Kanban system represents a practice already contained in PM concepts such as SAFe even at the higher project level,[31] which can also be found in practical implementation, albeit only sporadically so far.[32] This has already been discussed in the context of the flow principle in Section 1.2.3 and in detail in Section 4.5.2. Ultimately, this does indeed conceal a *paradigm shift in resource management* for a project portfolio: The philosophy of "Bring People to Projects!", i.e. staffing activated projects as closely as possible to demand, is giving way to the approach of "Bring Projects to People!", i.e. handling possible projects according to capacity. Such an approach can significantly facilitate classic resource management, which repeatedly leads to complications in companies. As in the production Kanban system, a fundamentally self-controlling system emerges with regard to the use of resources.

8.5.7 Perfection in the PPM

In Lean Management, striving for perfection refers to the efficiency of the processes carried out. Ultimately, the aim is to avoid waste of any kind or at least to reduce it as much as possible. The effectiveness of the processes is implicitly included, because the customer benefit, as the guiding criterion for process design, is the benchmark. Process results that are not accepted by the process customer lead inevitably to unnecessary rework, which in turn reduces process efficiency.

How customer expectations are optimally fulfilled is discussed in Section 8.5.2 (relationship between customer benefits and value creation processes). There perfection is expressed by attributes such as optimal, timely or appropriate. As is known from quality management, it is not the maximum possible quality that is perfect, but the quality for which the customer is willing to pay for, i.e. the required quality. The *Kano model,* known from marketing, describes this with the fulfilment of *basic and performance factors* according to customer expectations, but also formulates enthusiasm factors that ensure the desired customer loyalty and unique selling points/propositions.[33] The extent to which this approach can and should be applied in the more internally oriented PPM is certainly dependent on the corporate culture, in particular the internal services mentality.

---

30 see Schadwinkel 2018.
31 see Mathis 2016.
32 see Komus et al. 2020.
33 see e.g. Jochem 2019, pp. 57–61.

8.6   Preliminary concluding remarks on Lean-Agile PPM

The renowned benchmarking studies on multi-project management of the Universities of Technology in Berlin and Darmstadt regularly identify the key success factors of PPM (processes and decision-making culture).[34] These can often be recognised as the application of Lean Management practices (without being described as such there). As an example, the top performers of PPM monitor their portfolio more frequently and systematically and react more quickly to changing conditions, i.e. adjust the portfolio at short notice to avoid undesirable developments. Lean PPM therefore promises, through Lean Thinking and the approach of adapting proven Lean Management principles and practices, and by incorporating the concepts of Lean PM, a helpful further development of PPM.[35]

Lean-Agile PPM means – as does Lean PM – a more flexible design of the PPM system, i.e. especially its processes, adapted to the company's framework conditions. Rigid processes that are perceived as bureaucratic prove counterproductive in practice, because they regularly lead to disregard, as no benefit is associated with them or is not recognisable. Strict orientation towards the process customers and their benefit expectations creates the prerequisite for avoiding waste during implementation.

The following outstanding elements can be identified as characteristic contributions of Lean-Agile for a successful PPM:[36]

• Short project portfolio review cycles according to a company-wide cycle
• Flexible project budgeting via Strategic Buckets
• Acceptance through transparency, pragmatism, participation, CIP
• Reduction of project scope, duration, team size
• Capacity-oriented project management ("Bring projects to people!")

These core points represent a significant further development of the classical, stability-oriented PPM.

The investigation of the agilisation of PPM, as the possible antithesis to classic PPM, shows that due to the changed framework conditions for projects, results are often demanded more quickly for quite new, innovative project subjects (digitalisation) and with the simultaneous goal of improving project success rates. "One size fits all" will not apply in PPM either, because companies have different framework conditions (market, culture, etc.), or increasingly diverse projects in multimodal project landscapes (Scrum/agile, V-model/waterfall/sequential) etc. With the guideline of Lean Thinking, it can be possible to set a framework for efficient and effective PPM. The results of the multi-project management studies cited from the Universities in Berlin and Darmstadt show many elements of top performance in multi-project management that can be assigned to the idea of Lean-Agile PPM. Beyond the approaches described in this chapter, there is an exciting field for further conceptual and empirical research. But that's the story of another book.

---

34 see Gemünden 2019; Gemünden et al. 2016.
35 see Erne 2019; Hüsselmann et al. 2018; Pautsch/Steininger 2014; Seidl 2020.
36 see e.g. Certa/Albayrak 2018; Gemünden et al. 2016; Liechti 2019; SAFe 2020.

REFERENCES

Agile Business Consortium (Ed.) (2019). *Agile Project Management Handbook*. V2.

AHO (Ed.) (2014). *Untersuchungen zum Leistungsbild, zur Honorierung und zur Beauftragung von PM-Leistungen in der Bau- und Immobilienwirtschaft*. AHO Heft 9, Schriftenreihe des AHO. AHO Ausschuss der Ingenieurverbände und Ingenieurkammern für die Honorarordnung (e.V).

Aichele, C. & Schönberger, M. (2014). *IT-Projektmanagement: Effiziente Einführung in das Management von Projekten*. Springer.

Alt, R.,Auth, G. & Kögler, C. (2017). *Innovationsorientiertes IT-Management mit DevOps*. Springer.

Ambler, S. & Lines, M. (2020). *Introduction to Disciplined Agile Delivery* (2nd ed.). Project Management Institute.

Anderson, D. J. (2003). *Agile Management for Software Engineering. Applying the Theory of Constraints for Business Results*. Prentice Hall.

Andler, N. (2010). *Tools für Projektmanagement, Workshops und Consulting. Kompendium der wichtigsten Techniken und Methoden* (3rd ed.). Publicis.

Arndt, N. (2016). Gute Karten für die Thüringer Polizei. In *arcAKTUELL* 2/2016, S. 36–37. https://arcaktu ell.esridech.net/02-2016/#36, zuletzt geprüft am 24.07.2023

AWF (2007). Wertstrom in der Administration: Vom Ist- zum Sollzustand mit dem Wertstrom-Design. Arbeitsgemeinschaft für wirtschaftliche Fertigung. www.awf.de/downloads/artikel-dokumente/

AXELOS (Ed.) (2017). *Managing Successful Projects with PRINCE 2* (6th ed.). Stationery Office.

Ballard, G. (2012). *Target Value Design. Lean Construction Institute*. University of California. www.resea rchgate.net/publication/266501227_Target_Value_Design

Ballard, G. (2018). Das Last Planner System. In Fiedler, M. (Ed.), *Lean Construction – Das Managementhandbuch*, 121–135. Springer.

Ballard, G. & Howell, G. (2002). Lean project management. In *Building Research & Information*, 31(2), 119–133.

Barton, T., Herrmann, F., Meister, V., Müller, C. & Seel, C. (Ed.) (2017). Prozesse, Technologie, Anwendungen, Systeme und Management 2017. Fachtagung Angewandte Forschung in der Wirtschaftsinformatik: Tagungsband zur 30. AKWI-Jahrestagung vom 17.09.2017 bis 20.09.2017 an der Hochschule Aschaffenburg.

Barton, T., Herrmann, F., Meister, V., Müller, C. & Seel, C. (Ed.) (2018). Prozesse, Technologie, Anwendungen, Systeme und Management 2018. Fachtagung Angewandte Forschung in der Wirtschaftsinformatik: Tagungsband zur 31. AKWI-Jahrestagung vom 09.09.2018 bis 12.09.2018 an der Hochschule für Angewandte Wissenschaften Hamburg.

Bayer, P. (2010). System verstehen mit Cynafin. wandelweb.de 2010. www.wandelweb.de/blog/?p=962. Access date: 25.02.2021.

Beck, K. et al. (2001). Das agile Manifest. http://agilemanifesto.org/iso/de/manifesto.html. Access date: 25.02.2021.

Becker, J., Kugler, M. & Rosemann, M. (2005). *Prozessmanagement. Ein Leitfaden zur prozessorientierten Organisationsgestaltung* (5th ed.). Springer.

Becker, M. (2019). *Qualitätsmanagement: Die wesentlichen Grundlagen kompakt und verständlich.* Books on Demand.

Beer, S. (1994). *Decision and Control. The Meaning of Operational Research and Management Cybernetics.* Wiley.

Belvedere, V., Cuttaia, F., Rossi, M. & Stringhetti, L. (2019). Mapping wastes in complex projects for Lean Product Development. In *International Journal of Project Management*, 37 (3), 410–424.

Bertagnolli, F. (2018). *Lean Management. Einführung und Vertiefung in die japanische Management-Philosophie.* Springer.

Beyer, M. (2011). Bundesagentur für Arbeit schließt SAP-Projekt ab, computerwoche. www.computerwoche.de/a/bundesagentur-fuer-arbeit-schliesst-sap-projekt-ab,2364746. Access date: 25.02.2021.

Bicheno, J. (1998). *The Lean Toolbox. A Quick and Dirty Guide for Cost, Quality, Delivery, Design and Management.* Picsie Books.

Bicheno, J. |& Holweg, M. (2016). *The Lean Toolbox. A Handbook for Lean Transformation* (5th ed.). Picsie Books.

Blust, M. (2019). Methoden, Chancen und Risiken hybrider Projektmanagementvorgehensmodelle. In Linssen, O. et al. (Ed.), *Projektmanagement und Vorgehensmodelle 2019*, 69–82. Köllen-Verlag.

Blust, M. & Kan, E. (2019). *Vorgehensmodelle und Methoden im hybriden Projektmanagement. Eine empirische Studie. Landshuter Arbeitsberichte zur Wirtschaftsinformatik.* Hochschule Landshut.

BMF (Ed.) (2019). Stand der Doppelbesteuerungsabkommen und anderer Abkommen im Steuerbereich sowie der Abkommensverhandlungen am 1. Januar 2019. Rundschreiben. Bundesministerium der Finanzen.

BMFSFJ (Ed.) (2013). FAQ: Aufgaben von Arbeitsagentur und Jobcenter, Bundesministerium für Familie, Senioren, Frauen und Jugend. www.perspektive-wiedereinstieg.de/Inhalte/DE/Wiedereinstieg/Wiedereinstieg_konkret/Beratung_vor_Ort/faq_aufgaben_von_arbeitsagentur_und_jobcenter.html?view=pdf. Access date: 25.02.2021.

Boehm, B. W. & Turner, R. (2009). *Balancing Agility and Discipline. A Guide for the Perplexed* (7th ed.). Addison-Wesley.

Bouchard, S. (2017). *Lean Robotics. A Guide to Making Robots Work in Your Factory.* Bouchard.

Bowes, J. (2015). Kanban vs Scrumvs XP – an Agile comparison. https://manifesto.co.uk/kanban-vs-scrum-vs-xp-an-agile-comparison. Access date: 25.02.2021.

BPUG (2014). Verleihung des PRINCE2 Best Practice Awards 2014 auf dem BPUG Kongress. Best Practice User Group Deutschland e.V.

Brandenburg, J. et al. (2020). Der BER im 3D-Modell. Deswegen wurde 14 Jahre lang gebaut, Tagesspiegel. https://interaktiv.tagesspiegel.de/lab/flughafen-ber-in-3d/?utm_source=pocket-newtab-global-de-DE. Access date: 25.02.2021.

Brecht-Hadraschek, B. (2014). Projektmanagement-Zertifizierungen im Vergleich, Teil 1 & 2. In *projektmagazin*, 9/10, 4–24.

Brehm, L., Feldmüller, D. & Rieke, T. (2017). Konfiguration des hybriden Projektmanagements für die Entwicklung technischer, physischer Produkte. In Barton, T. et al. (Ed.), *Prozesse, Technologie, Anwendungen, Systeme und Management 2017*, 30–39. Fachtagung Angewandte Forschung in der Wirtschaftsinformatik: Tagungsband zur 30. AKWI-Jahrestagung vom 17.09.2017 bis 20.09.2017 an der Hochschule Aschaffenburg.

Brenner, J. (2018). *Lean Administration. Verschwendung in Büros erkennen, analysieren und beseitigen.* Hanser.

Brodbeck, F. (2016). Ergebnisse der GLOBE-Studie. In Brodbeck, F. et al., *Internationale Führung: das GLOBE-Brevier in der Praxis*, 87–165. Springer.

Brodbeck, F., Kirchler, E. & Woschée, R. (Ed.) (2016). *Internationale Führung*. Die Wirtschaftspsychologie.

Bürmann, R. &Hüsselmann, C. (2008). ERP-gestützte Personalwirtschaft in der Bundesverwaltung. In *Innovative Verwaltung* 5, 34–36.

Busse, M. (2017). Implementing Lean Management. Ein ganzheitliches Vorgehensmodell zur nachhaltigen Implementierung des Lean Managements in KMU (Dissertation). Brandenburgische TU Cottbus-Senftenberg.

Büttgen, M. & Fabricius, G. (Ed.) (n.d.). Planungsverhalten im Projektmanagement. www.gpm-ipma. de/fileadmin/user_upload/GPM/Know-How/Ergebnisbericht_Studie_Planungsverhalten.pdf. Access date: 25.02.2021.

Capgemini (Ed.) (2006). *Veränderungen erfolgreich gestalten – Change-Management 2005*. Capgemini.

Certa, S. & Albayrak, C. (2018). Eine hybride Vorgehensweise zur IT-Projektportfolioplanung. In Mikusz et al. (Eds.), *Projektmanagement und Vorgehensmodelle 2018*, 135–146. Gesellschaft für Informatik.

Cockburn, A. (2002). *Agile Software Development*. Addison-Wesley.

Colledge, B. (2005). Relational Contracting – Creating Value Beyond the Project. In *Lean Construction Journal*, 2, 30–45.

Constantine, J. (2001). Methodological Agility. In *Software Development*, 9(6), 67–69.

CrossCulture (Ed.) (n.d.). The Lewis Model. Richard Lewis Communications Ltd. www.crossculture.com/about-us/the-model. Access date: 25.02.2021.

Dathe, T., Helmold, M. & Hummel, F. (2019). *Erfolgreiche Verhandlungen: Best-in-Class Empfehlungen für den Verhandlungsdurchbruch*. Springer.

Decker, H. (2015). Großprojekte sind zum Scheitern verurteilt. FAZ Frankfurter Allgemeine Zeitung. www.faz.net/aktuell/wirtschaft/unternehmen/studie-grossprojekte-wie-der-ber-zum-scheitern-verurteilt-13827732.html. Access date: 25.02.2021.

Deming, E.W. (1982). *Quality, Productivity, and Competitive Position*. Massachusetts Institute of Technology.

Deutsche Gesellschaft für Projektmanagement (GPM) (Ed.) (2014). Empirische PMO Studie 2013/14, in Kooperation mit der HfWU Hochschule für Wirtschaft und Umwelt Nürtingen-Geislingen.

Deutsche Gesellschaft für Projektmanagement (GPM) (Ed.) (2017). Individual Competence Baseline für Projektmanagement, Version 4.0, German version.

Deutsche Gesellschaft für Projektmanagement (GPM) (Ed.) (2019). Kompetenzbasiertes Projektmanagement (PM4): Handbuch für Praxis und Weiterbildung im Projektmanagement.

Deutsche Gesellschaft für Qualität e.V. (Ed.) (2012). FMEA – Fehlermöglichkeits- und Einflussanalyse. DGQ-Band 13–11 (5th ed.).

Diebold, P. & Simon, F. (2019). Compliance: Umgang mit dem agilen Feind? In Linssen et al. (Ed.), *Projektmanagement und Vorgehensmodelle 2019*, 45–55. Gesellschaft für Informatik.

DIN (2009a). Projektmanagement – Projektmanagementsysteme – Teil 2: Prozesse, Prozessmodell (DIN 69901-2). DIN.

DIN (2009b). Projektmanagement – Projektmanagementsysteme – Teil 5: Begriffe (DIN 69901-5). DIN.

DIN (2015). Qualitätsmanagementsystem: Grundlagen und Begriffe, (DIN EN ISO 9000). DIN.

Drexl, N., Hans, S. & Käck, S. (2002). Hauptseminar Analyse von Softwarefehlern – Softwarefehler in der Logistik am Beispiel des Denver International Airport Gepäcktransportsystems. Technische Universität München. www5.in.tum.de/lehre/seminare/semsoft/unterlagen_02/denver/website. Access date: 25.02.2020.

Eilmann, p. et al. (2011). Interessengruppen/Interessierte Parteien. In Gessler, M./Deutsche Gesellschaft für Projektmanagement: Kompetenzbasiertes Projektmanagment (4th ed.), 1. Gessler, M. & GPM.

Erne, R. (2019). *Lean Project Management – Wie man den Lean-Gedanken im Projektmanagement einsetzen kann*. Springer.

Eube, A. (2016). Warum tragen einflussreiche Männer Einheitslook? WELT. www.welt.de/icon/article15 1608437/Warum-tragen-einflussreiche-Maenner-Einheitslook.html. Access date: 25.02.2021.

European Association of Business Process Management (EABPM) (Ed.) (2014). *Business Process Management. BPM Common Body of Knowledge – BPM CBoK, Version 3.0. Leitfaden zum Geschäftsprozessmanagement* (2nd ed.). Gießen.

Fahr, P., Gonzalez Arteta, M., Huber, M. (2019). SAP S/4HANA. BearingPoints Projekt-Erfahrung. Presentation TH Mittelhessen. Friedberg.

Felchlin, J. (2020). 3 Erfolgsfaktoren für agiles Portfoliomanagement. Quartalsweise Planung, aktive Auftraggeber und Project Tailoring. In *projektmagazin*, 64–71.

Feldmüller, D. (2018). Konfiguration des hybriden Projektmanagements nach Nutzenbetrachtungen. In Barton et al., *Prozesse, Technologie, Anwendungen, Systeme und Management*, 31, 177–186. AKWI-Jahrestagung.

Feldmüller, D. & Sticherling, N. (2016). Agile Methoden in der Entwicklung mechatronischer Produkte. In *Projektmanagement aktuell* 2, 14–22.

Felkai, R. & Beiderwieden, A. (2011). *Projektmanagement für technische Projekte: Ein prozessorientierter Leitfaden für die Praxis*. Springer.

Fiedler, M. (2018). *Lean Construction – Das Managementhandbuch. Agile Methoden und Lean Management im Bauwesen*. Springer.

Fleig, J. (2019). Mit Change-Management wird die Unternehmensstrategie umgesetzt. www.business-wissen.de/hb/was-ist-change-management-oder-veraenderungsmanagement. Access date: 25.02.2021.

Fleischmann, A., Schmidt, W., Stary, C., Obermeier, S. & Börger, E. (2011). *Subjektorientiertes Prozessmanagement. Mitarbeiter einbinden, Motivation und Prozessakzeptanz steigern*. Hanser.

Freund, J. & Rücker, B. (2014). *Praxishandbuch BPMN 2.0* (4th ed.), Hanser.

Frick, A., Schoper, Y., Röschlein, R. & Seidl, J. (2019). Projektdesign. In *GPM*, 1004–1037.

Gemünden, H.-G. (2019). *Success Factors of Project Portfolio Management. The Essence from 15 Years of Empirical Research*. TU Berlin/BI Norwegian Business School.

Gemünden, H.G. et al. (2011). *5. Multiprojektmanagement-Benchmarking-Studie*. TU Berlin.

Gemünden, H.G. et al. (2013). *6. Multiprojektmanagement-Benchmarking-Studie. Allgemeiner Abschlussbericht*. TU Berlin.

Gemünden, H.-G. & Kock, A. et al. (2016). *Erfolgsfaktoren im MPM: Ergebnisse der 7. Multiprojektmanagement-Benchmarking-Studie*. TU Darmstadt/TU.

Gessler, M. (2012). Projektarten. In Gessler, M. (Ed), *Kompetenzbasiertes Projektmanagement (PM3)*. Handbuch für die Projektarbeit, Qualifizierung und Zertifizierung auf Basis der IPMA Competence Baseline Version 3.0., 43–51. Deutsche Gesellschaft für Projektmanagement.

Gessler, M. (Ed.) (2011). *Kompetenzbasiertes Projektmanagement*. 8. Ed., Band 1. Deutsche Gesellschaft für Projektmanagement.

Gessler, M. & Deutsche Gesellschaft für Projektmanagement (GPM) (Ed.) (2012). *Kompetenzbasiertes Projektmanagement (PM3): Handbuch für Projektarbeit, Qualifizierung und Zertifizierung* (5th ed.). Deutsche Gesellschaft für Projektmanagement.

Gessler, M. & Sebe-Opfermann, A. (2013). Die Logik des Gelingens: Heuristiken im Projektmanagement. In Wald et al. (Ed.) (2013), *Advanced Project Management* 3, 65–76. Deutsche Gesellschaft für Projektmanagement.

Goldratt, E. (1990). *What Is This Thing Called Theory of Constraints and How Should It Be Implemented?* North River Press.

Goll, J. & Hommel, D. (2015). *Mit Scrum zum gewünschten System*. Springer.

GOOGLE Trends (2020) Suchbegriff „komplex". https://trends.google.de/trends/explore?date=all&geo=DE&q=komplex. Access date: 17.07.2020.

Gorecki, P. & Pautsch, P. (2013). *Lean Management. Auf den Spuren des Erfolges der Managementphilosophie von Toyota und Co.* (3rd ed). Hanser.

Grote, S. & Goyk, R. (2018). *Führungsinstrumente aus dem Silicon Valley: Konzepte und Kompetenzen*. Springer.

Grundlach, C. & Jochem, R. (2015). *Praxishandbuch Six Sigma, Fehler vermeiden, Prozesse verbessern, Kosten senken* (2nd ed.). Symposion Publishing.

Habermann, F. (n.d.). Agiler Populismus oder gute Sache? Das Beispiel der Stacey-Matrix. https://overt hefence.com.de/agiler-populismus-oder-gute-sache-das-beispiel-der-stacey-matrix. Access date: 25.02.2021.

Haemer, N. (2019). Lean Project Controlling. Übertragung des Lean-Gedankens auf das Einzel- und Multiprojektcontrolling (Masterthesis). TH Mittelhessen, PPM Labor.

Haghsheno, S., Schmitz, N. & Budau, M. (2018). Mehrparteienvereinbarungen auf Basis der Theorie relationaler Verträge. Ein Beitrag zur Lösung von Problemen konventioneller Projektabwicklungsformen bei komplexen Bauvorhaben? In Schwerdtner, P. & Kessel, T. Tagungsband zum 29. BBB-Assistententreffen. TU Braunschweig, 75–84.

Hall, E. T. (1990). *The Silent Language* (Reissue). Knopf Doubleday.

Hanser, E., Mikusz, M. & Fazal-Baqaie, M. (Ed.) (2013). Vorgehensmodelle – Anspruch und Wirklichkeit. 20. Tagung der Fachgruppe Vorgehensmodelle im Fachgebiet Wirtschaftsinformatik (WI-VM). Lecture Notes in Informatics (LNI) – Proceedings P-224. Gesellschaft für Informatik (GI).

Haric, P. (n.d.). Management. In Gabler Wirtschaftslexikon. https://wirtschaftslexikon.gabler.de/definit ion/management-37609. Access date: 25.02.2021.

Häusling, A., Römer, E. & Zeppenfeld, N. (2018). *Praxisbuch Agilität. Tools für Personal- und Organisationsentwicklung.* Haufe.

Hedeman, B. & Seegers, R. (2012). *PRINCE2® 2009 Edition. Das Taschenbuch* (5th ed.). Van Haren Publishing.

Heidemann, A. (2011). Kooperative Projektabwicklung im Bauwesen unter der Berücksichtigung von Lean-Prinzipien. Entwicklung eines Lean-Projektabwicklungssystems. Internationale Untersuchungen im Hinblick auf die Umsetzung und Anwendbarkeit in Deutschland. (Dissertation). Reihe F, Forschung, 68. Karlsruher Institut für Technologie. TU Karlsruhe.

Heinen-Konschak, E. & Brendle, B. (2017). Mit Effectuation Projekte im Ungewissen meistern. Strategien erfolgreicher Unternehmer im VUCA-Umfeld. In projektmagazin. www.projektmagazin.de/artikel/ mit-effectuation-projekte-im-ungewissen-meistern-teil-1_1121087. Access date: 25.02.2021.

Herbig, N. (2015). *Lean Dictionary*. Books on Demand.

Herneck, C. & Kneuper, R. (2011). *Prozesse verbessern mit CMMI for Services. Ein Praxisleitfaden mit Fallstudien.* Dpunkt.

Heylighen, F. & Joslyn, C. (2001). The Law of Requisite Variety. In Principia Cybernetica Web 2001. http://pespmc1.vub.ac.be/REQVAR.html. Access date: 25.02.2021.

Hills, M. D. (2002). Kluckhohn and Strodtbeck's Values Orientation Theory. Online Readings. In *Psychology and Culture*, 4(4). https://doi.org/10.9707/2307-0919.1040

Hinz, O. & Poczynek, J. (2011). Wider die zunehmende Verdosung des Projektmanagements. In *Organisationsentwicklung*, (1), 72–76.

Hofstede, G. (1997). *Lokales Denken, globales Handeln. Kulturen, Zusammenarbeit und Management.* Dt. Taschenbuch-Verlag.

Hofstede Insights (Ed.) (n.d.). Compare Countries. www.hofstede-insights.com/product/compare-countr ies. Access date: 25.02.2021.

Höhn, R., Höppner, S. & Rausch, A. (2008). *Das V-Modell XT. Anwendungen, Werkzeuge, Standards.* Springer.

Hope, J. & Fraser, R. (2003). *Beyond Budgeting: Wie sich Manager aus der jährlichen Budgetierungsfalle befreien können.* Schäffer-Poeschel.

Hüsselmann, C. (2014). Agilität im Auftragnehmer-Auftraggeber-Spannungsfeld. Mit hybridem Projektansatz zur Win-Win-Situation. In *Projektmanagement Aktuell*, 2014, (1), 38–42.

Hüsselmann, C. (2020). *Das Unified Project Management Framework. Ein kompakter Prozessrahmen für Projekte.* Wissenschaft & Praxis.

Hüsselmann, C. & Hemmann, T. (2006). Prozessorientierte Einführung von SAP R/3 HR in einer Bundesverwaltung. In Kruppke, H. et al. (Ed.), *Human Capital*, 143–161. Springer.

Hüsselmann, C., Leyendecker, B. & Heymann, M. (2018). Lean Project Management. Entwicklung eines Ansatzes zur Harmonisierung agiler und plangetriebener Projektansätze. WI-[Report] Nr. 004. TH Mittelhessen. Friedberg.

Hüsselmann, C., Litzenberger, R., Schick, N. & Spannenberger, L. (2019). Sensitivitätsanalyse von Nutzwerten. Systematische Betrachtungen zur Nutzwertanalyse. WI-[Report] 009. TH Mittelhessen. Friedberg.

Hüsselmann, C. & Seidl, J. (Ed.) (2015). *Multiprojektmanagement. Herausforderungen und Best Practices.* Symposion Publishing.

Imai, M. (2002). *KAIZEN. Der Schlüssel zum Erfolg im Wettbewerb* (2nd ed.). Econ Ullstein List Verlag.

Jenny, B. (2014). *Projektmanagement: Das Wissen für eine erfolgreiche Karriere* (4th ed.). df Hochschulvlg.

Jochem, R. (2019). *Was kostet Qualität? – Wirtschaftlichkeit von Qualität ermitteln* (2nd ed.). Hanser.

Kammerer, S., Lang, M. & Amberg, M. (Ed.) (2012). *IT-Projektmanagement-Methoden: Best Practices von Scrum bis PRINCE2.* Symposion Publishing.

Kerzner, H. (2003). *Projektmanagement. Ein systemorientierter Ansatz zur Planung und Steuerung.* MITP.

Kirchgeorg, M. (n.d.). Kunde. In Gabler-Wirtschaftslexikon. https://wirtschaftslexikon.gabler.de/definit ion/kunde-37108/wikipedia. Access date: 25.02.2021.

Kirchhof, M. & Kraft, B. (2020). Agile und klassische Methoden im Projekt passend kombinieren. Hybrides Vorgehensmodell. In *projektmagazin* (Ed.), Spotlight Hybrides Projektmanagement. Das richtige Vorgehen für Ihre Projekte finden, 9, 14–24. Berlab Media GmbH.

Kloss, R. (2019). Adding Value to Project Management – The Magic Triangle Meets the „Cultural" Iceberg. In Stangl-Meseke, M. et al. (Ed.), *Practical wisdom and diversity. Aligning Insights, Virtues and Value*, 205–218. Springer.

Klotz, M. & Marx, p. (2018). Projektmanagement-Normen und -Standards. SIMAT Arbeitspapiere, Nr. 10-18-033. Hochschule Stralsund.

Komus, A et al. (2020). *Status Quo (Scaled) Agile, Studie.* Hochschule Koblenz.

Komus, A. (2020). Agil-klassische Mischformen – neue Chancen und Herausforderungen für PMOs und Unternehmen. Selektiv, hybrid, bimodal? Teil 1 & 2. In *projektmagazin* (Ed.) Spotlight Hybrides Projektmanagement. Das richtige Vorgehen für Ihre Projekte finden, 9. 80–93. Berlab Media GmbH.

Komus, A. et al. (2018). *Status Quo PEP, Studie.* Hochschule Koblenz.

Komus, A. & Kuberg, M. (2017). Study Status Quo Agile. www.status-quo-agile.net. Access date: 25.02.2021.

Komus, A., Simon, C. & Müller, W. (2016). *Multitasking im Projektmanagement. Status Quo und Potentiale Studie.* Hochschule Koblenz & Vistem GmbH.

Korn, H.-P. (2013). Das „agile" Vorgehen: Neuer Wein in alte Schläuche – oder ein „Déjà-vu"? In Hanser, E. et al. (Ed.), *Vorgehensmodelle – Anspruch und Wirklichkeit*, 109–132. Gessekschaft für Informatik.

Kotter, J. P. (2011). *Leading Change: Wie Sie Ihr Unternehmen in acht Schritten erfolgreich verändern.* Vahlen.

Kroker, M. (2018). Die lange Liste schwieriger und gefloppter SAP-Projekte, Wirtschaftswoche. www. wiwo.de/unternehmen/it/haribo-lidl-deutsche-post-und-co-die-lange-liste-schwieriger-und-gefloppter-sap-projekte/23771296.html. Access date: 25.02.2021.

Krüger, W., Bach, N. & (2014). *Excellence in Change – Wege zur strategischen Erneuerung* (5th ed.). Springer.

Kruppke, H., Otto, M. & Gontard, M. (Ed.) (2006). *Human Capital Management – Personalprozesse erfolgreich managen.* Springer.

Kurtz, K. & Sauer, J. (2018). Auswirkungen des Einsatzes hybrider Methoden auf die Projektsteuerung. In Mikusz, M. et al. (Ed.), *Projektmanagement und Vorgehensmodelle 2018*, 73–83. Bonn/Düsseldorf.

Kuster, J., Bachmann, C., Huber, E., Hubmann, M., Lippmann, R., Schneider, E., Schneider, P., Witschi, U. & Wüst, R. (2019). *Handbuch Projektmanagement. Agil – klassisch – Hybrid* (4th ed.). Springer.

Kuster, J., Huber, E., Lippmann, R., Schmid, A., Schneider, E., Witschi, U. & Wüst, R. (2011). *Handbuch Projektmanagement* (3rd ed.). Springer.

Kuwert, G (2019). Was macht Prozessmanagement aus? Presentation Kongress CPO@BPM&O Digitalisierung & Automatisierung.

Lauer, T. (2019). *Change Management. Grundlagen und Erfolgsfaktoren* (3rd ed.). Springer.

LCI (Ed.) (2017). *Lean Project Delivery Glossary*. Lean Construction Institute.

Leach, L.P. (2005). *Lean Project Management. Eight Principles for Success*. Advanced Projects, Inc.

Lean Construction Blog (n.d.). An Introduction to Set-Based Design. https://leanconstructionblog.com/introduction-to-set-based-design.html. Access date: 25.02.2021.

Leffingwell, D (2018*). SAFe 4.5 Reference Guide: Scaled Agile Framework for Lean Enterprises* (2nd ed.). Addison-Wesley.

Leopold, K. (2017). *Kanban in der Praxis. Vom Teamfokus zur Wertschöpfung*. Hanser.

LeSS (n.d.). Large-Scale Scrum. The LeSS Company B.V. https://less.works/de. Access date: 25.02.2021.

Lewis, R. D. (2006). *When Cultures Collide: Leading Across Cultures. A Major New Edition of the Global Guide* (3rd ed.). Nicholas Brealey Publishing.

Leyendecker, B. (2015). Six Sigma in administrativen Prozessen und Dienstleistung. In Grundlach, C./Jochem, R. (Ed.), *Praxishandbuch Six Sigma*, (2nd ed.), 133–148. Symposion Publishing.

Liechti, M. (2019). Eine Umsetzung von „Lean Portfolio Management" bei der Schweizerischen Mobiliar Versicherung. Presentation PM Forum 2019 der Deutschen Gesellschaft für Projektmanagement.

Liker, J. K. (2007). *Der Toyota-Weg. 14 Managementprinzipien des weltweit erfolgreichsten Automobilkonzerns* (3rd ed.). FinanzBuch.

Lingnau, V., Mayer, A. & Schönbohm, A. (2004). Beyond Budgeting. Notwendige Kulturrevolution für Unternehmen und Controller? *Beiträge zur Controlling-Forschung*, Bd. 6. TU Kaiserslautern.

Linssen, O., Mikusz, M., Volland, A., Yigitbas, E., Engstler, M., Fazal-Baqaie, M. & Kuhrmann, M. (Ed.) (2019). Projektmanagement und Vorgehensmodelle 2019. Neue Vorgehensmodelle in Projekte – Führung, Kulturen und Infrastrukturen im Wandel. Gesellschaft für Informatik. www.researchgate.net/publication/336738762_Projektmanagement_und_Vorgehensmodelle_2019_Neue_Vorgehensmodelle_in_Projekten_-_Fuhrung_Kulturen_und_Infrastrukturen_im_Wandel

Lock, D. & Wagner, R. (Ed.) (2019). *The Handbook of Project Portfolio Management*. Routledge.

Madauss, B.-J. (2017). *Projektmanagement*. Springer.

Marx, S. & Klotz, M. (2020). Earned-Value-Analyse: Einführung und Beispiele, SIMAT Arbeitspapiere, No. 03-20-036. Hochschule Stralsund.

Mathis, C. (2018). *SAFe – Das Scaled Agile Framework. Lean und Agile in großen Unternehmen skalieren* (2nd ed.). dpunkt.

McKendrick, J. (2015). How Amazon handles a new software deployment every second. www.zdnet.com/article/how-amazon-handles-a-new-software-deployment-every-second. Access date: 25.02.2020.

Mields, J. & Kuczynski, I. (2020). Nudging – ein neuer Weg zur Senkung von Unfallrisiken. Whitepaper. BG ETEM Berufsgenossenschaft Energie Textil Elektro Medienerzeugnisse.

Mikusz, M., Volland, A., Engstler, M., Fazal-Baqaie, M., Hanser, E. & Linssen, O. (Ed.) (2018). *Projektmanagement und Vorgehensmodelle 2018, PVM Tagung 2018. Der Einfluss der Digitalisierung auf Projektmanagementmethoden und Entwicklungsprozesse*. Gesellschaft für Informatik.

Mintzberg, H. (1991). *Mintzberg über Management. Führung und Organisation, Mythos und Realität*. Springer.

Oesterreich, B. (2013). Wir brauchen mehr Unsicherheit, mehr Komplexität und mehr Selbstorganisation. In Wald, A. et al. (Ed.), *Advanced Project Management*, (3), 79–93. GPM.

Office of Government Commerce (Ed.) (2017)/ *Managing Successful Projects with PRINCE2* (6th ed.). Stationery Office.

Ohno T. (2013). *Das Toyota-Produktionssystem* (3rd ed.). Campus.

Opelt, A., Gloger, B., Pfarl, W. & Mittermayr, R. (2018*). Der agile Festpreis. Leitfaden für wirklich erfolgreiche IT-Projekt-Verträge* (3rd ed.). Hanser.

Ottmann, R., Pfeiffer, A. & Schelle, H. (2008). *Projektmanager* (3rd ed.). GPM.

Parkinson, C. (1955). Parkinson's Law. In *The Economist* Nr. 5856, 11/1955, Bd. 177, 635–637.

Patzak, G. & Rattay, G. (2014). *Projektmanagement: Projekte, Projektportfolios, Programme und projektorientierte Unternehmen* (6th ed.). Linde.

Pautsch, P. & Steininger, p. (2014). *Lean Project Management. Projekte exzellent umsetzen.* Hanser.

Peipe, S (2015). *Crashkurs Projektmanagement* (6th ed.). HAufe.

Planview (n.d.). What is Set-Based Design. www.planview.com/de/resources/articles/lkdc-set-based-des ign. Access date: 25.02.2021.

PMI (2004). *A Guide to the Project Management Body of Knowledge.* PMBOK® Guide (3rd ed.). PMI.

PMI (2005). *Organizational Project Management Maturity Model: OPM3 Knowledge Foundation.* PMI.

PMI (2017). *A Guide to the Project Management Body of Knowledge.* PMBOK® Guide (6th ed.). PMI.

PMI (Ed.) (2018). PMI Pulse of the Profession. Success in Disruptive Times. www.pmi.org/-/media/pmi/ documents/public/pdf/learning/thought-leadership/pulse/pulse-of-the-profession-2018.pdf?sc_lang_t emp=en. Access date: 25.02.2021.

Pommeranz, I. (2011). Komplexitätsbewältigung im Multiprojektmanagement: Die Handlungsperspektive der Multiprojektleiter (Dissertation). Universität Augsburg.

Pommeranz, I. (2011). Taxonomie der Komplexität. In Pommeranz, I. 47–61.

Poppendieck, M., Poppendieck, T. (2003). *Lean Software Development. An Agile Toolkit.* Addison-Wesley.

Praxisframework (n.d.). Stacey matrix. www.praxisframework.org/en/library/stacey-matrix. Access date: 25.02.2021.

Preuß, N. (2014). Projektmanagementleistungen in der Bau- und Immobilienwirtschaft (4th ed.). Schriftenreihe des AHO, Bd. 9. AHO Ausschuss der Verbände und Kammern der Ingenieure und Architekten für die Honorarordnung e.V.

projektmagazin (Ed.) (2014). PM-Standards und -Zertifizierungen im Überblick. Spotlight – Eine themenspezifische Zusammenstellung von Fachartikeln aus dem Projekt Magazin. www.projektmaga zin.de. Access date: 03.01.2018 (online nicht mehr verfügbar).

projektmagazin (Ed.) (2020). Spotlight Hybrides Projektmanagement. Das richtige Vorgehen für Ihre Projekte finden, Bd. 09. Berleb Media GmbH.

PROMIDIS (Ed.) (2015). Instrument Makigami. PROMIDIS Handlungsleitfaden.

Reinhard, M. (2015). Portfoliomanagement im Kontext komplexer dynamischer Systeme. In Hüsselmann, C. & Seidl, J. (Ed.), *Multiprojektmanagement*, 145–191. Springer.

Reiss, M. (2018). Simplexity – Strategien für das Projektmanagement. In *Projektmanagement aktuell*, 29(3), 40–46.

Rietiker, p. (2013). Ergebnisse einer etwas anderen Projektmanagement-Studie – „Misserfolgsfaktoren in der Projektarbeit". In GPM Blog. www.gpm-blog.de/misserfolgsfaktoren-projektmanagement-studie. Access date: 25.02.2021.

Robinson, A. (2015). The (Internet of Things) IOT Supply Chain Benefits Now Coming Clearer. https:// cerasis.com/iot-supply-chain/. Access date: 25.02.2021.

Romeike, F. (2018). *Risikomanagement.* Springer.

Rother, M. (2010). Projekte, Programme, Portfolios – P3O, Leitfaden für eine gut strukturierte Projektlandschaft. www.projektmagazin.de/artikel/p3o-leitfaden-fuer-eine-gut-strukturierte-projekt landschaft_7291. Access date: 25.02.2021.

Royce, W. (1970). Managing the Development of Large Software Systems. In Proceedings of IEEE WESCON 26, 328–338.

Rüegg-Stürm, J. (2005). *Das neue St. Galler Management-Modell. Grundkategorien einer integrierten Managementlehre. Der HSG-Ansatz* (2nd ed.). Haupt.

SAFe (2020). SAFe® for Lean Enterprises 5.0. Scaled Agile, Inc. www.scaledagileframework.com. Access date: 25.02.2021.

Sandhaus, G., Berg, B. & Knott, P. (2014). *Hybride Softwareentwicklung. Das Beste aus klassischen und agilen Methoden in einem Modell vereint.* Springer.

Schadwinkel, A. (2018). Rettungsaktion in Thailand: Was wir wissen – und was nicht. ZEIT. www.zeit.de/wissen/2018-07/rettungsaktion-thailand-hoehle-tham-luang-taucher. Access date: 25.02.2021.

Schäfer, F. (2017). *Passgenaues Projektmanagement. Mit Augenmaß zum Projekterfolg.* Consult.

Scheer, A.-W. (Ed.) (2017). *Projektportfoliomanagement. Aktuelle Bedeutung und zukünftige Entwicklung.* Scheer Studie.

Schelle, H. (2013). Entwicklungsgeschichte und Trends im Projektmanagement. In. Wagner, R./Grau, N. (Ed.), *Basiswissen Projektmanagement*, 109–123. Symposion Publishing.

Scheller, T. (2017). *Auf dem Weg zur agilen Organisation: Wie Sie Ihr Unternehmen dynamischer, flexibler und leistungsfähiger gestalten.* Vahlen.

Schmitz, A. (2018). Gartner: 10 Technologietrends für 2018. https://news.sap.com/germany/2018/01/gartner-technologietrends-2018/?source=social-de-nl_kw4-xing-2017_News2Share-blog-newscenter-10-technologietrends-2018&campaigncode=CRM-DE17-SOC-SMC_MA01. Access date: 25.02.2021.

Schnichels-Fahrbach, L./Munz, A. (2016). Agiles Projektportfoliomanagement. In Wagner, R. (Ed.), *Erfolgreiches Projektportfoliomanagement*, 215–243. Symposion Publishing.

Schott, E. & Campana, C. (Ed.). (2005). *Strategisches Projektmanagement.* Springer.

Schott, E. & Wick, M. (2005). Change-Management. In Schott, E./Campana, C. (2005), *Strategisches Projektmanagement*, 195–231. Springer.

Schütte, R., Seufert, S. & Wulfert, T. (2019). Das Wertbeitragscontrolling als Anreicherung bestehender Vorgehensmodelle des Software Engineerings. In Linssen, O. et al. (Ed.), *Projektmanagement und Vorgehensmodelle 2019*, 111–126. Gesselschaft für Informatik.

Schwaber, K. & Sutherland, J. (2017). Der Scrum Guide. Der gültige Leitfaden für Scrum: Die Spielregeln, deutsche Ausgabe. www.scrumguides.org. Access date: 03.03.2021.

Schwaiger, M. (2003). Der Student als Kunde – eine empirische Analyse der Zufriedenheit Münchner BWL-Studenten mit ihrem Studium. In *IHF Beiträge zur Hochschulforschung*, 25(1), 32–62.

Schwerdtner, P. &Kessel, T. (Ed.) (2018). Tagungsband zum 29. BBB-Assistententreffen. Fachkongress der wissenschaftlichen Mitarbeiter der Bereiche Bauwirtschaft | Baubetrieb | Bauverfahrenstechnik. Zentrum für Bau- und Infrastrukturmanagement. 06. –08.06.2018. TU Braunschweig.

Seel, C. & Timinger, H. (2017). Ein adaptives Vorgehensmodell für hybrides Projektmanagement. In Barton, T. et al. (Ed.). *Prozesse, Technologie, Anwendungen, Systeme und Management 2017*, 20–29. Hochschule Aschaffenburg.

Seidl, J. (04.09.2020). Die Zukunft des Projekt(portfolio)managements ist lean! Presentation 11. PM-Tag Rhein-Ruhr Projektmanagement der Zukunft, Dortmund/online.

Seidl, J. (2011). *Multiprojektmanagement.* Springer.

Shenhar, A. J. & Dvir, D. (2007). *Reinventing Project Management. The Diamond Approach to Successful Growth And Innovation.* Harvard Business Review Press.

Software Engineering Institute (2006). CMMI for Development, Version 1.2. Carnegie Mellon University.

Sonntag, A. (2015). Das Instrument Makigami. www.inf.uni-hamburg.de/de/inst/ab/itmc/research/completed/promidis/instrumente/makigami. Access date: 25.02.2021.

SPIEGEL Wirtschaft (n.d.). Der Tag, an dem die Wallstreet kollabierte. www.spiegel.de/wirtschaft/unternehmen/15-september-2008-der-tag-an-dem-die-wall-street-kollabierte-a-648261.html. Access date: 25.02.2021.

Sprenger, R. K. (2014). *Mythos Motivation: Wege aus einer Sackgasse* (20th ed.). Campus.

Špundak, M. (2014). Mixed Agile/Traditional Project Management Methodology – Reality or Illusion? In *Procedia – Social and Behavioral Sciences* 119, 939–948.

Stacey, R. D. (2007). *Strategic Management and Organisational Dynamics. The Challenge of Complexity* (5th ed.). Financial Times/Prentice Hall.

Staehle, W. H. (Ed.) (1991). *Handbuch Management. Die 24 Rollen der exzellenten Führungskraft.* Gabler.

Staehle, W. H., Conrad, P., Sydow, J. (2014). *Management. Eine verhaltenswissenschaftliche Perspektive* (8th ed.). Vahlen.

Stalk, G. & Hout, T. (1990). *Competing Against Time.* Free Press.

Standish Group (Ed.) (2015). CHAOS Report 2015. The Standish Group International, Inc.

Stangel-Meseke, M., Boven, C., Braun, G., Habisch, A., Scherle, N. & Ihlenburg, F. (Ed.) (2019). *Practical Wisdom and Diversity: Aligning Insights, Virtues and Values.* Springer.

Steeger, O. (2014). Notruf 110 – jetzt hat Erfurt übernommen! Pünktlichkeit dank Projektmanagement: Die Einsatzleitzentrale wurde „scharf geschaltet". In Projektmanagement aktuell, 25(2) 16–24. https://docplayer.org/11889425-Notruf-110-jetzt-hat-erfurt-uebernommen.html. Access date: 03.03.2021.

Sterrer, C. (2014). *Das Geheimnis erfolgreicher Projekte – Kritische Erfolgsfaktoren im Projektmanagement – Was Führungskräfte wissen müssen.* Springer.

Süß, R. (2016). Standards für das Projektportfoliomanagement. In Wagner, R. (Ed.), *Erfolgreiches Projektportfoliomanagement,* 123–139. Symposion Publishing.

t2informatik (Ed.) (n.d.a). Das V-Modell XT spezifisch zuschneiden. https://t2informatik.de/wissen-kompakt/tailoring. Access date: 25.02.2021.

t2informatik (Ed.) (n.d.b). Was ist VUCA und welche Strategie leitet sich daraus ab? https://t2informatik.de/wissen-kompakt/vuka. Access date: 25.02.2021.

t3n (Ed.) (n.d.). devops. https://t3n.de/tag/devops. Access date: 25.02.2021.

Techt, U./Lörz, H. (2015). *Critical Chain: Beschleunigen Sie Ihr Projektmanagement* (3rd ed.). GPM.

The Standish Group (Ed.) (2018). NEW Chaos Report (2018). www.standishgroup.com/news/37. Access date: 25.02.2021.

Tie, M. (2020). Lean im Projektmanagement. Webinar, REFA/IMPULS.

Timinger, H. (2017). *Modernes Projektmanagement. Mit traditionellem, agilem und hybridem Vorgehen zum Erfolg.* Wiley.

Timinger, H., Paukner, M. & Seel, C. (2018) Projektparameter für das Tailoring hybrider Projektmanagementvorgehensmodelle. In Barton, T. et al. (Ed.), *Prozesse, Technologie, Anwendungen, Systeme und Management 2018,* 166–176. Hochschule für Angewandte Wissenschaften Hamburg.

Trompenaars, F. & Hampden-Turner, C. (1997). *Riding the Waves of Culture. Understanding Cultural Diversity in Business.* Nicholas Brealey Publishing.

Ulrich, H. (Ed.) (1973). *Unternehmensplanung: Bericht von der wissenschaftlichen Tagung der Hochschullehrer für Betriebswirtschaft in Augsburg vom 12. bis 16. Juni 1973.* Springer.

Untereiner, G. (2013). *Frankreich lohnt sich. Handbuch für den erfolgreichen Export. Schwerpunkte: Vertriebsorganisation, Firmengründung und Firmenerwerb.* CIRAC.

VersionOne Inc. (2017). 11th annual State Of Agile Report. https://explore.versionone.com/state-of-agile/versionone-11th-annual-state-of-agile-report-2. Access date: 03.04.2018.

Vogelsang, K. & Olberding, J. (2007). Projektmanagement in KMU: Eine Sammlung von Best Practices. www.projektmagazin.de/artikel/projektmanagement-kmu-eine-sammlung-von-best-practices_7023. Access date: 25.02.2021.

Wagner, R. (Ed.) (2016). *Erfolgreiches Projektportfoliomanagement. Wie Sie Projektportfolios systematisch gestalten und steuern.* Symposion Publishing.

Wagner, R. & Grau, N. (Ed.) (2013). *Basiswissen Projektmanagement – Projekte steuern und erfolgreich beenden.* Symposion Publishing.

Wald, A., Mayer, T.-L., Wagner, R. & Schneider, C. (Eds.) (2013). *Advanced Project Management, 3. Komplexität. Dynamik. Unsicherheit.* GPM.

Wald, A., Schoper, Y. et al. (2015). *Makroökonomische Vermessung der Projekttätigkeit in Deutschland.* GPM.

Wendler, R. (2009). Reifegradmodelle für das IT-Projektmanagement. *Dresdner Beiträge zur Wirtschaftsinformatik,* 53(9). TU Dresden.

Wiendahl, H. (2014). *Betriebsorganisation für Ingenieure* (8th ed.). Hanser.

Wilhelm, H. (2019). *Agiles und klassisches Projektmanagement im Change-Projekt „Manufacturing Execution System (MES)" – Welche Methoden führen zum Projekterfolg? (Bachelorthesis).* TH Mittelhessen, PPM Labor.

Wirfs-Brock, R. (2011). Agile Architecture Myths #2 Architecture Decisions Should Be Made At the Last Responsible Moment. http://wirfs-brock.com/blog/2011/01/18/agile-architecture-myths-2-architecture-decisions-should-be-made-at-the-last-responsible-moment. Access date: 25.02.2021.

Wojewoda, S. & Hastle, S. (2015). Standish Group 2015 Chaos Report – Q&A with Jennifer Lynch. www.infoq.com/articles/standish-chaos-2015. Access date: 25.02.2021.

Womack, J. P. & Jones, D. T. (2013). *Lean Thinking. Ballast abwerfen, Unternehmensgewinn steigern* (3rd ed.). Campus.

Womack, J. P., Jones, D. T. & Roos, D. (1991). *Die zweite Revolution in der Autoindustrie. Konsequenzen aus der weltweiten Studie des Massachusetts Institute of Technology.* Campus Verlag.

Zaninelli, S. (1995). Sechs-Stufen-Modell eines interkulturellen „integrativen Trainings". Dargestellt am Beispiel des Moduls: Arbeits- und Organisationsstile. 4.5.2.1. In Deutscher Wirtschaftsdienst (Ed.), *Handbuch für Personalentwicklung und Training* 29, 1–27.

Zell, H. (2017). *Projektmanagement – lernen, lehren und für die Praxis* (7th ed.). Books on Demand.

INDEX

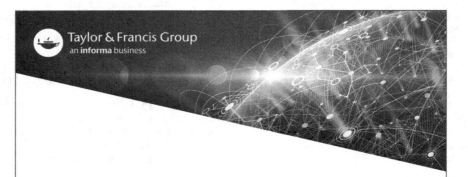